WILDFLOWER WONDERS
THE 50 BEST WILDFLOWER SITES IN THE WORLD

WILDFLOWER WONDERS
THE 50 BEST WILDFLOWER SITES IN THE WORLD

BOB GIBBONS
FOREWORD BY RICHARD MABEY

BLOOMSBURY
LONDON · NEW DELHI · NEW YORK · SYDNEY

First published in 2011 by New Holland Publishers

This paperback edition published in 2014 by Bloomsbury Publishing Plc

Copyright © 2011 in text and photographs: Bob Gibbons and other contributors as listed below or in the site accounts.

ISBN 978-1-4729-0982-4

Bloomsbury Publishing Plc, 50 Bedford Square, London WC1B 3DP

www.bloomsbury.com

Bloomsbury is a trademark of Bloomsbury Publishing Plc

Bloomsbury Publishing, London, New Delhi, New York and Sydney

A CIP catalogue record for this book is available from the British Library.

Publisher: Nigel Redman

Editor: Beth Lucas

Designer: Nicola Liddiard

Printed in China by C&C Offset Co. Ltd

10 9 8 7 6 5 4 3 2 1

All images by Bob Gibbons except the following: Ros Salter (page 107, both); Ian Green (page 129, top); Chris Gardner (pages 129, bottom, 131, 132, 133 and 137); and Chris Grey-Wilson (pages 134, 135 and 136).

Contents

FOREWORD BY RICHARD MABEY

Everywhere there are miniature floral gems – tufts of viridian wood sorrel on mossy stumps, fair fields full of poppies – and everywhere great tracts of subdued plant diversity on grazed-back grasslands. The rain forests have their solitary, inaccessible wonders, and a single stand of English bluebells shows how even a monoculture can generate something magically luminous. Such places provide powerful but confined experiences. But here and there, and more especially here and then, because favourable climatic spells are what count most of all, these factors can come together, and spark a flowering which is both extravagantly vast, hugely diverse, and uncompromisingly visible. A spectacle, in fact.

It is these moments, and the places where they most often occur, that the distinguished photographer Bob Gibbons celebrates in this collection of portraits of the world's floweriest places, from the interlapping sweeps of laceflowers, Hillside Daisies and San Joaquin Blazing Stars of the Carrizo Plains of California, to the austere limestone rockscapes of the Taurus Mountains in Turkey. This Anatolian mountainscape provides contrasting nuances of floral profusion, intense, distilled, perfectly placed: giant snowdrops amongst the white rocks, winter aconites spangling the scree at the edge of the snowline, as far as the eye can see.

There is of course an element of subjectivity in judging what constitutes superlative floweriness. Sheer tonal power, variety, form, colour blendings – all have their champions. Bob Gibbons' lifetime in the field has developed in him a special skill for capturing not just the character and loveliness of individual blooms but the complicated and related details of a flowering landscape, and it has earned him the right to present his own vision as a benchmark. His eye is for the sharp and high coloured detail of Renaissance paintings.

His vote for the most flowery place in the world goes to a real scene that could have come from a Medieval Book of Hours, the kaleidoscopic grasslands of Mount Rainier in Washington State. Gibbons gives as much to the scene as he gratefully takes from it: the early visit to catch the dew, which makes the lupins look as if they have been new – minted; the fantastical partridge foot spikes presiding like May Poles amongst the multitudes, the whole tapestry set like a warm benediction against the backdrop of Mount Rainier's glacial slopes. This is where the long processes of evolution, and a human imagination long-tutored in their ways, come together in a moment of epiphany, that says so much about our world and our common understanding of natural beauty.

INTRODUCTION

This book is a celebration of the world of wildflowers. We hear so much about what we are destroying – with good reason, as we have lost so much in recent decades – but my aim here has been to find and show some of the most beautiful and flowery places that are still left in the world.

It's no simple matter to define 'most flowery'. Essentially, this is a personal selection of the places that I find to be most flowery, spectacular and inviting to visit. The broad criteria I have used are spectacular beauty and diversity coupled with reasonable accessibility. Many areas of tropical rainforest, especially those in South America and South-east Asia are extraordinarily diverse botanically, but their structure and their lack of a short explosive flower season means that they are not a great spectacle for the non-specialist, and they have been excluded. Similarly, there are many wonderfully flowery places whose spectacle depends largely on one or two species – British bluebell woods, heathland, or a mountain meadow full of wild daffodils, for example, and I have excluded these, except where they appear as part of larger sites. I have also excluded very small sites – most of the places in the book warrant a visit of at least a couple of days, often a week or more, at the right season.

Many of these places are visited by specialist tour companies, and I have given contact details for the best of those that I know of, as well as some useful websites and other contacts in the Useful Contacts section (see page 179).

These are all fabulous places. Enjoy them and respect them – they are some of the finest treasures that we have.

Above: The gorgeous blue flowers of King-of-the-Alps in Engadin, Switzerland.

Below: Spring flowers in South Africa's Namaqua National Park.

WHY ARE SOME PLACES SO FLOWERY?

Above: A lovely clump of sticky primula at 3000m in Switzerland.

Opposite above: An astonishing display of mountain flowers near Crested Butte, USA.

Opposite below: Masses of Partridge Foot and other flowers on Mount Rainier, USA.

There is no simple answer to the question of why some parts of the world are particularly flowery whilst others are not. A number of features have to come together, but two major influences explain why some areas are astonishingly flowery: the actions of nature and man.

Much of the world's vegetation has been heavily influenced by man over very long periods. In many cases, this has led to a degradation of the diversity of the vegetation, particularly where there is arable cultivation, heavy grazing or forestry (and, of course, development, which leaves little that is natural in its wake). Vast areas of the world would undoubtedly be more flower-rich were it not for the hand of man, especially in the most populous and developed nations. In some cases, though, man's intervention may have made sites more flowery, not simply where there are managed nature reserves or parks, but in many other ways. Traditional agriculture, especially pastoral agriculture carried out without modern fertilizers and pesticides, can lead to an intensely flowery landscape, such as the Transylvanian grasslands of Romania or the machair of the Outer Hebrides, both farmed for generations in a sustainable and low-intensity way. A more subtle change results from long-term pastoralism at and around mountain treelines, which extends the area of alpine pasture. Such high wild areas are rarely farmed intensively, thanks to the remote and difficult terrain, and these high grasslands can be wonderfully flowery places, as seen in Turkey or the Alps.

The main factor behind the most spectacular flower-rich places is timing. A small window of time into which most flowering is concentrated produces the most dramatic and satisfying displays. This rules out most tropical sites where the seasons are not sufficiently distinct to produce a major flowering peak. The two situations that produce the best masses of flowers are hot dry summers preceded by damp or cold winters, and heavy winter snow-cover which does not melt until midsummer.

The hot dry summer preceded by a damp winter is epitomized by the Mediterranean climates of the world (which occur all around the Mediterranean Sea, as well as in California, western Australia, South Africa, and Chile). Flowering varies both in timing and intensity according to latitude, altitude and other local variations, but typically there is a strong burst from late March through April in the northern hemisphere and late August through September in the southern hemisphere, before the heat and drought of summer begins to bite. Where the climate is close to semi-desert or desert – such as in far southern California or northern Namaqualand – flowering can be intense and short-lived. Here it is strongly influenced by winter

rainfall, which can easily fall below the level needed for good flowers. In such places some years are astounding while others are disappointing.

Many parts of the world receive heavy winter snowfall, but by no means all of these – even where natural habitat remains – produce wonderful summer displays of flowers. The main determining factor seems to be the shortness of the succeeding summer, and this is most likely to be produced by very heavy winter snow that does not clear until June or July. On Mount Rainier in the Cascade Mountains, and in parts of the Rockies, an exceptional intensity of summer flowering is squeezed between the snow and the deteriorating weather of early autumn. A similar situation can occur where winter snow is followed quite quickly by a dry summer, such as on the Swedish island of Öland, where flowering is concentrated into a short period in May and early June before the shallow limestone soils dry out.

In all these places, you can feel the rush to complete flowering, pollination and fruiting, and this is matched by the frantic activity of animals and birds in the habitat. In the more unpredictable semi-desert areas, the flower colours are often of extraordinary intensity, evolved to attract limited numbers of pollinators, as insect populations cannot build up as quickly as plant populations in these circumstances. In addition to their bright colours, flowers here have evolved other special ways of attracting insects, such as the fake beetles of the Beetle Daisy in Namaqualand.

Surprisingly, the underlying geology of an area seems to have relatively little effect on flowering intensity. It is probably true that limestones – so beloved of botanists for their tendency to produce a plethora of rare flowers, especially orchids – predominate, but there are also plenty of good sites on granite, volcanic rocks, sandstone, schist, serpentines, and even pure sand.

Whatever the reasons for each site's special nature, it is a privilege to be at any of them when they reach their peak of flowering – all are beautiful places.

The Burren

Opposite above: A beautiful clump of Shrubby Cinquefoil on the limestone pavement at Mullagh Mor.

Opposite below: A typical stony field in the Burren uplands, dominated by Mountain Avens and Early Purple Orchids.

There is something particularly special about the Burren. It's a strange and wonderful place, full of contradictions. The white limestone landscape is at once both stark and harsh yet intimate and beguiling, packed with ancient stone monuments and criss-crossed by old stone walls and drove roads. At first sight, it looks barren, with barely a plant to be seen, yet as soon as you walk a few yards from the road you are suddenly surrounded by flowers, growing from the most unlikely positions. The landscape is a tapestry of secret corners, hidden groves, old buildings and plant treasures. Lakes come and go almost overnight and woods are no higher than a man's head; and where else in a cool temperate climate do cattle go up into the mountains to graze in winter and come down to the lowlands in summer?

The core of the Burren is a great block of Carboniferous limestone running down to the coast of County Clare on the southern shore of Galway Bay, and extending outwards to encompass the wonderful Aran Islands not far offshore, covering about 350 km² in all. The whole area was once heavily glaciated, leaving a harsh bare rocky landscape full of cracks. Huge areas of almost bare limestone pavement, with big flat-topped blocks (clints) are separated by deep fissures known as grykes or scailps. The cracks are home to an abundance of flowers, often woodland plants, while the clints are only flowery where a thin soil has been able to develop. Where fields have been enclosed and perhaps cleared of rock, there are masses of nodding yellow Cowslips and other flowers.

Two features make the area really special. First, the flowers that do occur grow in an extraordinary and wonderful abundance – drifts of Early Purple Orchids, clusters of intensely blue Spring Gentians, long cracks filled with rose-red Bloody Cranesbills, beautiful scattered clumps of yellow-flowered Shrubby Cinquefoil, thousands of purple Large-flowered Butterworts cascading down a damp grassy slope and hillsides dripping with mats of white Mountain Avens. Secondly, plants grow in unique combinations: Mediterranean plants such as Dense-flowered Orchid or Hoary Rock Rose grow just centimetres away from arctic specialities such as Mountain Avens or Norwegian Sandwort, or mountain plants like Spring Gentian. Nowhere else do these combinations occur.

The reasons for the special nature of the Burren are complex and hard to define. It's a combination of the underlying porous calcareous rock, the glaciation, the long history of colonization, clearance and settlement by man, and the mild yet harsh Atlantic climate, almost frost-free and with relatively little variation in

temperature through the year. Rainfall is higher in winter, and it is this that leads to the cattle being grazed on the higher pastures (winterages) in winter, then brought down in summer. This absence of grazing in spring and summer leads to a wonderful explosion of flowers on the higher pastures, though this system is threatened by current changes in farming, especially those driven by the EU trend towards homogeneity.

The best areas are around Black Head and Poulsallagh along the coast facing the Aran Islands and the small protected national park area near Mullagh Mor, but almost anywhere is worth a visit in late May.

14

The machair of the Outer Hebrides

INFORMATION

Location | Best seen on the western coasts of the Outer Hebrides islands, especially North and South Uist.

Reasons to go | Unlike anywhere else. Spectacular displays of flowers; lots of birds; wonderfully evocative wild scenery.

Timing | Peak flowering in late June and July, but plenty to see from mid-April onwards.

Protected status | The best areas are protected by SSSI (Site of Special Scientific Interest) status and international designations, and there are a few managed nature reserves.

Britain's north-western coasts are wild, windy and unashamedly beautiful places, often exposed to the full force of Atlantic gales sweeping in from the west. In places, the sand of the beaches is startlingly white, a mixture of quartz and the sea-ground remains of shells. Because of the high shell content, this sand can be highly calcareous, and in the most exposed conditions it blows inland, modifying the peaty soil to make it more lime-rich, drier, and sometimes more fertile. For centuries, local crofters have casually cultivated these areas, planting occasional crops of potatoes or cereals, then leaving them fallow for a few years. In fallow periods the cultivated patches become astonishingly flowery, a blaze of rainbow colours. A wonderful mixture including annuals, biennals and perennials, both common and rare, lasts for months in the cool damp climate.

In some definitions, the Gaelic word machair is used solely to describe these areas of occasionally cultivated land, though often the definition encompasses the more stable dune grasslands on the seaward side and the less disturbed peaty soils just inland. The total habitat includes a wonderful variety of flowers, from Heath Spotted Orchid, sundews and Bog Asphodel at the damp acidic end of the spectrum, through to Moonwort and Common Restharrow at the most calcareous

Right: A Ringed plover in breeding plumage amongst flowery machair.

Opposite: Intensely flowery machair with Red Clover, Ladies' Bedstraw and other flowers at Stilligarry with Ben More beyond.

dry end, on stable dunes. But it is the cultivated intermediate area which is so special, for it is unlike any other habitat. It is a unique product of an extreme climate and physical conditions, combined with an unusual form of land use.

No two patches of machair are quite the same, but most will contain a rich permutation including some of the following species as dominant colours: the rich red of honey-scented Red Clover, the massed yellow spikes of Lady's Bedstraw, white mats of White Clover, the lovely blue-purple inflorescences of Tufted Vetch (normally a scrambling plant, but here reduced to dense mats), deep purple Self-heal, the untidy pink flowers of Ragged Robin, or the yellow mats of Bird's Foot Trefoil. Sometimes, a whole patch will be dominated by the orange flowers of Corn

Right: Lovely species-rich grassland machair at Howmore on South Uist.

Opposite: Wherever machair has been recently cultivated, masses of cornfield weeds such as these Corn Marigolds appear.

Marigold, scarlet Field Poppy, or yellow and white mayweeds. In places, the lovely blue and yellow flowers of Heartsease are abundant, while elsewhere there may be masses of Hebridean Spotted Orchids, Hebridean Marsh Orchids, Early Marsh Orchids in their attractive wine-red form, or other distinctive rarer perennials. It is an astonishing and vibrant mixture, enhanced by the sweep of a beautiful empty white sand beach and the dark brooding mountains inland.

The machair is also wonderfully rich in breeding birds, with densities as high as anywhere in the world for a few species, such as Dunlin and Ringed Plover. There are also more Great Yellow Bumblebees than anywhere else, Grey and Common Seals offshore, and Common Otters hunting along the seaweedy shorelines. It's an irresistible and magical combination.

Machair occurs in many places along the western coasts of Eire, Ulster and Scotland, but it is at its best in the far north-western islands of Scotland, wherever a flat or gently sloping area is exposed to the full stretch of the Atlantic, without islands or headlands to lessen the force of the wind and waves. It reaches its zenith on the outer hebridean islands of Barra, South Uist, North Uist, Harris and Lewis, with key sites at Balranald around the lovely RSPB reserve, and almost anywhere down the western coasts of the Uists.

The Lizard Peninsula, Cornwall

INFORMATION

Location | Southern Cornwall, south of Helston and Falmouth.

Reasons to go | Lovely displays of spring coastal flowers in unspoilt scenery; many rarities; marvellous late summer heathlands. Coastal birds, good rockpools.

Timing | Particularly flowery between mid-April and mid-May, then again in late July and August, but interesting lower plants and spectacular coastal scenery can be seen all year.

Protected status | Many parts of the peninsula are protected as nature reserves or under National Trust ownership, and further areas are designated as SSSIs

Hundreds of places around the western coasts of England and Wales have wonderful displays of flowers in April and May. What makes the Lizard extra special is that it not only has stunning displays of widespread species, but it is also home to an exceptional number of rare plants, often occurring in remarkable quantity.

The Lizard Peninsula is the southernmost part of mainland Britain, with the warmest climate in the country on average and relatively low rainfall. It is also rather remarkable geologically. First, there is much the largest exposure of serpentine rock in Britain, covering over 60 km^2. Serpentine rock gives rise to rather infertile, shallow soil, which tends not to be intensively farmed, and supports a range of flowers that are rare or absent elsewhere. Secondly, the geology is complex and varied, with schists, gneiss, gabbro, gravels, shales and other rocks providing suitable conditions for a wide range of flowers.

To see the best displays of coastal flowers in spring, head for Mullion Cove and walk south, or go to Lizard Point. In a good year, the drifts of colour in April and early May can be astonishing. Swathes of Bluebells (more commonly a woodland plant, but quite frequent on south-western clifftops), pale yellow primroses, pink clumps of Thrift, white Ox-eye Daisies and Sea Campion, clumps of Kidney Vetch, usually in the normal pale yellow form, but sometimes red or white, Red Campion and Wild Carrot are all here in abundance. Some of the less common plants may be present in remarkable quantities – intense yellow mats of Hairy Greenweed, lovely masses of Green-winged Orchids, pale blue Spring Squills or the deep magenta Bloody Cranesbill. In recent years, there has been more and more colour from invasive plants such as Three-cornered Leek and several members of the mesembryanthemum family – all attractive plants, but apparently ousting some of the native flora here.

The area around Kynance Cove and southwards towards Lizard Point is particularly rich botanically, with many uncommon species. A 19th century botanist, the Rev. C. Johns is remembered for performing his famous hat trick here, throwing down his hat and counting six different species of clover under it, as well as many other flowers. This remains unquestionably the best site in Britain for clovers, with 14 species in one valley. Many other specialities on this stretch of coast include prostrate Gorse and Common Broom, Wild Chives, Thyme Broomrape, Wild Asparagus, and Wild Chamomile. A quite different strand in The Lizard tapestry is the extensive heathland on the plateau, much of it protected and managed as a national nature reserve. There are flowers of interest here through most of the spring and summer, but the heaths come into their own in late summer. Easily the largest (and virtually the only) population of Cornish Heath in Britain is quite magnificent, stretching away into the distance. It often grows with Bell Heather, Ling and Western Gorse, creating a wonderful pink, purple and yellow mosaic of colour.

Outside the normal flowering season, spectacular coastal scenery and an abundance of lichens and other lower plants make the Lizard worth a visit at almost any time of year.

Left: Coastal flowers on the cliffs above Mullion Cove, with Thrift, Sea Campion, Bluebells and other flowers in May.

Right: A lovely mixture of Spring Squill and Kidney Vetch in clifftop turf.

Abisko National Park

INFORMATION

Location | Far north-western Sweden, close to the Norwegian border, between Kiruna and Narvik.

Reasons to go | One of the best places for displays of Arctic tundra flowers, with many northern specialities; a good cross-section of Arctic birds, as well as Elk and Reindeer; Saami culture.

Timing | Rather variable according to snow levels, but the best time is almost always between mid-June and mid-July.

Protected status | The core of the area is the 7,700 ha Abisko National Park, but the area of interest extends widely to the north and west.

Abisko lies well north of the Arctic Circle and can be extremely cold. It offers a real taste of the Arctic and its tundra flora, yet it does so under relatively civilized conditions with easy access by road or rail (or even by air to nearby Kiruna). Northwards from here, particularly in Norway, are vast areas of fine unspoilt tundra, but the advantage of Abisko – in addition to its easy access – is a particularly rich suite of less common flowers in addition to the widespread Arctic species.

Winters here are very cold, but the area is the driest place in Sweden, with an average annual rainfall of only 300 mm. The park is in the rain shadow of mountains in all directions, and summer days here, apart from being extremely

long, are also frequently dry and sunny. Geologically, it is complex and varied, including shales and dolomite, with many of the rocks giving rise to calcareous soils.

The cold Arctic regions are never rich by more southerly standards, but Abisko has superb displays of beautiful flowers with many uncommon or rare species, and the dramatic Abiskojåkka Canyon has a wonderful range of special plants where it cuts through the hard calcareous schist rocks.

The calcareous heaths are beautifully flowery. This is the place for mats of Mountain Avens, Cowberry, Norwegian Wintergreen, the tiny Arctic Bellflower, intense blue Snow Gentian (one of the few annual plants to do well under these difficult conditions), several dwarf willows, some pretty little eyebrights and a number of orchids such as Fragrant, Small White and Frog Orchids. Two species of Cassiope can be found here, both covered with delicate little white or pinkish bell-shaped flowers. This is also the only Swedish site for the rare Arctic Orchid.

Opposite: The tiny shrubby Dwarf Cornel can grow in huge flowery masses here.

Below: At this latitude, the woodlands are open and filled with flowers such as Wood Cranesbill and Northern Wolfsbane.

The lower slopes of the calcareous mountains, such as Mount Njulla, support a surprisingly lush and colourful flora, dominated by Wood Cranesbill, tall stems of Garden Angelica, Globe Flower, Red Campion, buttercups, Mountain Sorrel, several lady's mantles, Water Avens and tall blue spikes of Northern Wolfsbane. In open birch woods, there may be a similar mixture, with the addition of glorious white carpets of Dwarf Cornel, and occasional rarities such as the elusive and capricious Ghost Orchid.

Higher on the mountains, the real tundra begins. Surprisingly, snow does not lie late here – the wind keeps the snow shallower than in the valleys, and early summer sun soon melts it. This is the signal for a brief explosion of flowers – the purplish-pink flowers of Arctic Rhododendron covering the dwarf bushes, cushions of Diapensia, covered with pretty white flowers, and dramatic mats of Purple Saxifrage, often flowering as soon as the snow goes. Also found here are pinkish-red Bog Rosemary, several yellow or white saxifrages, more cassiopes, Rose-root, several louseworts, and the lovely white Glacier Crowfoot, turning pink as it goes over. You are also quite likely to meet herds of Reindeer, and walk to the piping sounds of nesting Golden Plover or Dunlin.

The gorge cut by the main river as it hurtles down to the lake, though not spectacularly flowery, is home to a fascinating range of Arctic specialities, many of them rare elsewhere.

A nature centre, the Naturum, has excellent displays, guides and information about the park and surrounding area. In any exploration, don't be confined to the relatively small national park – the surrounding areas are equally fascinating.

24

Öland

Opposite: The strikingly beautiful Small Pasque Flower grows abundantly in the limestone pastures of Öland.

There is nowhere quite like the long thin island of Öland. It suffers from cold, snowy Scandinavian winters, yet has an almost Mediterranean summer climate; there are virtually no hills, yet it has a remarkably varied landscape; and its long history of settlement is constantly in evidence. It also has a rich flora, with over 1,000 species including a number of endemics.

Geologically, it is largely hard Ordovician limestone, scraped bare through the ice ages, and now mainly flat and open. Most of the central southern part of the island is occupied by a huge open plain, known as the Great Alvar or Stora Alvaret, where the limestone is never far from the surface and there are few trees or villages. The Great Alvar extends over 300 km² – the largest area of this type of habitat in Europe – and there are also many smaller areas of alvar, making up about a quarter of the island's surface area in total. This is the island's key area of botanical interest, but by no means the only one.

Öland has a long history of human settlement, and some wonderful prehistoric monuments and old villages, but it is a curiously disjointed history as the population has risen and fallen due to plague and wars. From the mid 16th century to the early 19th century, the whole island was a royal hunting forest, and local people were obliged to kill all predatory creatures, amputate a leg from any dog that might be able to hunt, and refrain from any hunting themselves. In the 1970s, a bridge from the mainland was finally built, and the area now prospers again, but its landscape and wildlife continue to be heavily influenced by the island's past.

A visit to the barer, rockier parts of one of the prime areas of alvar, such as those to the east of Vickleby, in late May or early June, might reveal masses of the pretty yellow endemic Öland Rock Rose, and the rather similar Hoary Rock Rose, growing with Meadow Saxifrage, Kidney Vetch in various colours, several cranesbills, some striking clumps of the lovely yellow Shrubby Cinquefoil, Dwarf Mouse Ear, Basil Thyme and several species of milkwort. Where the soil is a little deeper, the turf explodes into colour as Elder-flowered Orchids in both yellow and red forms (known locally as Adam and Eve for this reason) vie with Early Purple Orchid, Cowslip, Ox-eye Daisy, and even some very flowery dwarf forms of Blackthorn to form a splendid tapestry of colour. In a few places, there are wonderful stands of the early flowering Yellow Pheasant's Eye, perhaps growing with Pasque Flowers or the commoner Small Pasque Flower. In winter-damp areas, you can find swathes of the lovely pink Bird's Eye Primrose, often in its dwarf stemless form, with several violets, spotted orchids, water crowfoot and Marsh Thistles.

Elsewhere, the alvar may be more wooded with open clearings, such as around the wonderful 5th century fort at Ismantorp. Massed spikes of Military Orchids and Sword-leaved Helleborines draw the eye, but closer examination soon reveals clumps of yellow Viper's Grass, beautiful scented Lily-of-the-Valley, blue and pink milkworts, the curious greenish-flowered Swallow Wort and masses of frothy white Dropwort. Some lovely ancient woodlands, such as the oak woods at Halltorps Hage, are full of wonderful drifts of Wood Anemone, Yellow Wood Anemone, Hepatica, Common Lungwort, parasitic Toothwort and many orchids. Even the pine forests, in the far north of the island, can be flowery with cow wheats, helleborines, Twinflower, May Lily and Chickweed Wintergreen.

Opposite: One of the most extraordinary sights on Öland is the massed ranks of Early Purple and Elder-flowered Orchids.

Above: Sheets of Wood Anemones and Yellow Wood Anemones in Hazel and oak woodland at Halltorps Hage in April.

Osmussar southwards to Parnu

INFORMATION

Location | The western coast of Estonia and its offshore islands.

Reasons to go | Lovely unspoilt flowery habitats rich in orchids and many other species, often in astonishing quantity.

Timing | Good from mid-April until July, usually best mid-May to mid-June.

Protected status | Well-protected, with many nature reserves and national parks.

F lowery wooded meadows must be some of the most welcoming and sublime of all habitats, with their wonderful mixture of sunshine and shade filled with flowers, birds and insects. They exist in plenty in this unspoilt corner of Estonia. The glorious wooded meadow at Laelatu is said to hold the world record for density of plants (at 76 species per square metre, with a total species list of almost 500 higher plants), and is a most beautifully flowery place.

Underlying much of the west coast of Estonia and its islands, a layer of hard Carboniferous limestone supports a lovely mosaic of meadows, woodland and open alvar, richly covered by beautiful flowers. At its best, the open alvar is as flowery as that in Öland (p.24), with masses of stately pink Military Orchids, Green-winged Orchids, sheets of purple Wild Chives, nodding yellow Cowslips, pale Dog Violets, thousands of nodding Small Pasque Flowers and, perhaps most impressive of all here, vast dense beds of the beautiful pure white Snowdrop Windflower stretching away into the distance.

Both coniferous and deciduous forests may be flowery, depending on age and management. Dense beds of Angular and Common Solomon's Seals intermingle with huge quantities of fragrant Lily of the Valley, a sea of nodding plum-coloured Water Avens, patches of clear white Sword-leaved Helleborine, and masses of the distinctive Herb Paris. A little earlier, the same areas may be coloured blue, white and yellow with Hepatica, Wood Anemones and Yellow Anemones in vast drifts, dotted with the pinkish-white spikes of parasitic Toothwort. Perhaps the best of

Right: Part of a vast bed of fragrant Lily-of-the-Valley in woodland at Laelatu.

Opposite: The spring displays of Lady's Slipper Orchids in some of the wooded meadows are quite exceptional.

these woods are on Puhtu peninsula near Virtsu, and the fabulous Loode Oak Grove near Kuressaare; at the latter site, everything occurs in abundance, and the Lily-of-the Valleys are picked in industrial quantities by local people, with apparently little effect.

But the wooded meadows are the richest and most flowery habitats. Nowhere else have I seen so many Lady's Slipper Orchids, arguably the most beautiful of the terrestrial orchids, growing in huge clumps in grassy woodland glades. More open areas may be coloured yellow with Viper's Grass, or pink with Bird's Eye Primrose, or magenta-blue with Wood Cranesbill, while shadier patches have masses of blue and purple Spring Vetch as well as woodland species. Another special feature of these wooded meadows is that they are home to so much other wildlife, as Common Rosefinches sing from the tree tops, Icterine Warblers and Golden Orioles can be heard in the oaks, and dragonflies fly constantly around the clearings. You may even catch a glimpse of a grazing Elk or deer, while in the distance you can hear Common Cranes and Bitterns calling from the wetlands. These wooded meadows are idyllic places in almost every respect, except one: several million mosquitoes also love them.

It's possible to find good flowery sites throughout most of western Estonia, but some of the best are on the island of Osmussaar, still farmed traditionally by one family; in the complex of reserves around Virtsu; and in the big Mullutu-Loode Natura 2000 site just west of Kuressare on Saaremaa Island. All are wonderfully diverse species-rich areas in an unspoilt environment.

Opposite: Cowslips and Wood Forget-me-not make a glorious combination in May in open woodland on Saaremaa Island.

Above: Beautiful white Snowdrop Windflowers growing in massed ranks in grassland near Virtsu on the west coast.

The Vercors Mountains

One of the best days I have ever spent anywhere in the countryside of Europe was in the Vercors Mountains. We walked up the valley from St Michel les Portes (below the scarp on the east side of the range) towards the incredible peak of Mont Aiguille, passing through a series of wonderful meadows buzzing with insects, followed by beautiful beech woodland full of treasures like Lady's Slipper Orchids, Herb Paris and Lily-of-the-Valley; lunchtime found us in a warm flower-filled clearing where we saw a Golden Eagle fly overhead carrying a marmot (not so good for the marmot, but exciting for us). After lunch we climbed on over the shoulder of the mountain, amongst masses of alpine flowers, further great clumps of Lady's Slippers and a host of other orchids, before finally descending through the woods to Richardière and a welcome cold drink. The great thing about the Vercors is that there are many such places.

This spectacularly beautiful range of mountains is made up almost entirely of thick bands of Lower Cretaceous limestone tilted sharply towards the west, producing a long dramatic series of cliffs and peaks all down the eastern side. Tourists tend to pass it over in favour of the nearby Alps, leaving it surprisingly quiet. Above about 1,000 m, most of the meadows are strikingly full of flowers, vibrant with Wood Cranesbill, clovers, bedstraws, yellow rattles, knapweeds and Ox-eye Daisies, but given extra substance by many rare and beautiful flowers.

Right: Damp meadows below the western scarp cliffs have a glorious mixture of Pheasant's Eye Narcissi, Globeflowers, orchids, gentians and many other flowers.

Some fields are full of Pheasant's Eye Narcissus, others may be pink with Military Orchids or Early Purple Orchids; slightly damper meadows can be a mass of lovely lemon-yellow Globeflowers, while drier slopes are blue with Chalk Milkwort, matted blue globularias or yellow Horseshoe Vetch.

Some of the higher meadows, for example on the Col d'Arzelier or Col du Prayer, seem to have hardly any grass, dominated instead by a dozen or more species of orchid, blue Round-headed Rampion or Large Speedwell, the intense blue of Trumpet Gentians and Meadow Clary, and white spikes of St. Bernard's Lily. Open woods of Scots Pine, especially where they grow on limestone scree and downwash, can be sensationally flowery, with amazing groups of Lady's Slipper Orchids growing amongst many other orchids, lovely pink clumps of Round-leaved Restharrow, two species of Solomon's seal, several wintergreens, Swallow Wort, gromwells and bellflowers (to name but a few).

The very high pastures, on the high plateau or around the cliffs of the higher peaks like Mont Aiguille, may have more bare ground, but they are alive with orchids, gentians, milkworts, white Alpine Pasque Flowers and the lovely purple-blue Haller's Pasque Flower, yellow cinquefoils, Seguier's Buttercup and a range of more typical alpine species. Altogether, it's a beautiful, tranquil and astonishing place. There is plenty to occupy anyone in the Vercors for a week or more, though while in the area you might also consider the Dévoluy mountains to the south-east, or the Chartreuse Mountains to the north.

Top: Flowery hay meadows below Mont Aiguille. The upper slopes of the peak, below the cliffs, are also beautifully flowery.

Above: An appealing mixture of globularias, Trumpet Gentian, Horseshoe Vetch and a Lady Orchid on a limestone slope.

The Écrins National park

INFORMATION

Location | Eastern France, between the towns of Briançon and Gap.

Reasons to go | Lovely meadows and pastures amongst high granite and crystalline mountains with glaciers; over 1,800 species of flower including many that are rare, endemic or protected.

Timing | Best between late May and early July, peaking around early June in most years, later in the high valleys.

Protected status | The whole area falls within the huge Écrins National Park and is well-protected.

Opposite: It's hard to imagine a more flowery meadow than this one, dominated by Wood Cranesbill and Bistort, in the Upper Narreyroux Valley.

If you ever plan to go from Puy St Vincent up the Narreyroux Valley, or from Ailefroide up towards the Glacier Blanc to look at flowers in June, leave yourself plenty of time – these are seriously flowery valleys. All around the great bulk of the high Écrins, valleys carry foaming torrents away from the high peaks and glaciers, and they are all good for flowers. Within the core protected zone of the national park, opportunities for finding wonderful displays of flowers in early summer are endless.

The huge mountain range known as the Écrins is part of the Alps, yet somehow slightly separate from them. It lies to the west of the main range, separated by the deep valley of the Durance, and it stands alone as an almost circular massif with rivers draining out in all directions. The high peaks and most of the valleys are formed of granite, which produces a wonderful range of habitats for flowers, as it weathers readily into a deep rich soil, but remains too rocky for easy cultivation. In contrast to the nearby limestone mountains of the Vercors, granite mountains have much more surface water, more bogs and flushes, and usually more forest.

At lower levels, the Écrins are wooded or cultivated, with grasslands here and there. Above about 1,000 m, the meadows and pastures are almost all full of flowers, often astonishingly so. A typical high valley, such as that of the Entraigues or the Torrent d'Ailefroide has a lovely patchwork of flower meadows – sometimes fenced, sometimes open – woodlands, scrub, rocky areas with open flowery patches, big boulders, cliffs, and wet areas, perhaps where a side-stream spreads its water over the valley floor. Everywhere is flowery.

The best meadows are alive with colour and sounds. The intense pink Jove's Flower is abundant here, with several alpine bellflowers, the tall spikes of Yellow Gentian, masses of pinkish-red Martagon Lilies, and occasional Orange Lilies, Alpine Woundwort, the distinctive and beautiful blue-purple spikes of Meadow Clary, Wild Sainfoin, mulleins, and many more. Higher up, meadows are purple and pink with an extraordinary density of Bistort, Wood Cranesbill, at least four species of bellflower, Plane-leaved Buttercup, and the deep blue of Perennial Cornflowers.

On drier banks, there may be masses of stately Lizard Orchids, several types of pink, Yellow Woundwort, Musk Thistle, dark blue spikes of False Sainfoin, several blue scabiouses, and the magenta flowers of Tuberous Vetch. In slightly shadier places, the curious but beautiful pink or white spikes of Burning Bush can be found. This member of the citrus family is so-called because it produces a highly flammable gas that may burst into flame spontaneously, and certainly does if lit.

The shadier parts of rocks can be flowery too, with several species of house-leek and stonecrop, lovely Wood Pinks and clusters of white sandworts.

Along streams, or where springs emerge, there are often extensive boggy areas with crimson marsh orchids, Elder-flowered Orchids, Common and Heath Spotted Orchids, pink Bird's Eye Primrose, Grass of Parnassus and Cotton Grass. Above about 1,700 m there are many more alpine species such as gentians, rock jasmines, Edelweiss, mountain primulas and many other lovely species, but it is the mid-altitude areas that provide the best displays.

A wonderful bonus of coming to see flowers in the Écrins is the other wildlife here – about 140 species of butterfly, often in spectacular numbers, herds of Chamois and Ibex, and a long list of breeding birds which includes almost 40 pairs of Golden Eagles, Alpine Choughs and Tengmalm's Owl.

Not far away from the Écrins, east of Briançon and Guillestre, as far as the Italian border, the lovely Queyras valley is also full of little-visited flowery places.

Opposite: A beautiful, dew-covered Martagon Lily flowering in a high hay meadow.

Below: A wonderful stand of Yellow Gentians, with purple Meadow Clary and many other flowers in old pasture.

38 The Cévennes and Causses

INFORMATION

Location | Between the towns of Millau and Alès, stretching southwards towards Béziers and north as far as Mende.

Reasons to go | Almost too many to mention: wonderful flowers, with some outstanding displays and a huge range of species; a rich fauna of butterflies, other insects, and birds; beautiful scenery and fascinating medieval architecture.

Timing | Best between April and June, peaking around late May in most years.

Protected status | The best sites fall either within the Cévennes National Park or the enormous Grands Causses Natural Regional Park, covering 316,000 ha, though the latter by no means confers full protection.

Opposite: The gorgeous wine-red flowers of the Cévennes Pasque Flower in the limestone grasslands of the Causses.

There can be few things more pleasant in life than spending a few days around the limestone causses of southern France in May or early June. The scenery is spectacular, flowers burst forth from every angle, butterflies are everywhere – and of course, the food and wine are wonderful.

Around the busy town of Millau, on the river Tarn, is a huge Jurassic limestone plateau, deeply cut into dramatic gorges wherever a major river flows through it. This is wonderfully unspoilt countryside, too harsh for much intensive farming, yet mellow enough to support an astonishing diversity of plants and animals. The plateaux between the gorges lie at around 1,000 m, so winters are cold and hard, yet the dry hot summers are almost Mediterranean in their intensity. These high grassy plateaux are known locally as causses, and each block, subdivided by the gorges, has its own evocative name.

All the causses are lovely, each with a different character, but let's take one in particular to give a flavour of the area. Start at the lovely little twin village of Le Rozier-Peyreleau, where the gorges of the Tarn and the Jonte meet, and climb southwards out of the gorges on the impressive D29 (or up one of the many steep footpaths if you prefer). The north-facing slopes of the gorges are usually heavily wooded, in contrast to the more open and inhabited south-facing slopes, and you'll pass through patches of pine and oak woodland. Every so often, there is a grassy clearing, full of flowers and alive with dancing butterflies.

Blue Bugle seems to be more succesful here than anywhere else in its range, forming huge bright blue patches almost as far as the eye can see; White, Common and Hoary Rock Roses are all wonderfully abundant, interspersed with lovely blue or pink patches of Chalk or Common Milkworts, pale blue Aphyllanthes, looking almost like a flax though actually in the lily family, lovely pink cushions of Rock Soapwort wherever the soil is a little drier, or masses of blue Viper's Bugloss where there has been some disturbance. And then the orchids, in startling and varied abundance. Stately pink Military and Lady Orchids in unbelievable numbers, dark spikes of Early Purple and Green-winged Orchids, pretty little Burnt Orchids, Common Spotted Orchids in various colour forms, and many more. Enigmatic Fly Orchids, hard to spot but fascinatingly beautiful in close-up, are common, often growing with Bee Orchids, the endemic Aymonin's Orchid or the Small Spider Orchid – all members of the genus Ophrys, which have evolved to attract male bee and wasp pollinators who believe that the flowers are the females they are seeking. Clumps of pink Mountain Kidney Vetch, sprawling masses of a golden

drop, and many other lovely flowers can be seen on rocky limestone outcrops.

In dappled shade, there is a different selection of flowers. Elegant white spikes of Sword-leaved Helleborine jostle with remarkable quantities of the curious brown-flowered saprophytic Bird's Nest Orchid, as well as fragrant masses of Lily-of-the-Valley, Solomon's Seal, Bastard Balm (often in its deep pink form) amongst bushes of Snowy Mespil and Box. Deeper into the shade, there are blue hepaticas, delicate One-flowered Wintergreens and masses of the strange Yellow Bird's Nest. The mixture is similar on the plateau, where patches of grassland stretch away into the distance without hedges or fences. One part may have masses of blue-purple Pasque Flowers, another may have dwarf irises, while a third may be yellow with big clumps of Yellow Pheasant's Eye.

Just to the east of the causses, the more acid hills of the Cévennes provide a quite different experience. Although not so flowery, wonderful sights here include fields of Wild Daffodils and Pheasant's Eye Narcissus (and hybrids between the two), wild tulips in their reddish southern form, drifts of white Wood Anemones, Coral-root Bittercress and the related Five-leaved Bittercress, May-lilies and many more, often under lovely ancient pollarded beech trees.

Opposite: A group of vibrant Pyramidal Orchids stand out among the flowers of an old pasture in the Tarn Gorge.

Above: Wood Anemones and Wild Daffodils form a dense carpet in the higher more acidic grasslands of Mont Aigoual.

The central French Pyrenees

Location | Due south of Tarbes and Lourdes, reaching the Spanish border along the high peaks.

Reasons to go | Lovely glaciated high mountain scenery with abundant flowers in high pastures, cliffs and meadows. Over 2,500 species of flower, of which at least 200 are endemic. Rich and varied bird and mammal life.

Timing | Peak flowering from late May to July, usually best May through June, varying according to altitude and amounts of winter snow.

Protected status | Many of the main places of interest lie within the core area of the Pyrenees National Park, though a substantial number of meadows lie in the much larger buffer zone which is less well protected.

Opposite: Drifts of English Iris stretch away as far as the eye can see on the slopes of the Pic du Midi de Bigorre.

Heading south from Bordeaux or Toulouse, the great wall of the Pyrenees rises from the plains in front of you like a mirage, almost too high to be believed. Often enough, the tops will be shrouded in cloud, but on a clear day the sight of high snowy mountains stretching from horizon to horizon is unforgettable.

These are the most southerly mountains of the French mainland, forming a major social and ecological barrier between France and the Iberian peninsula. They are large and high enough to create their own weather, and isolated enough to have their own special flora and fauna, in some abundance. The central part of the Pyrenees has the highest peaks (see also p. 46) with a spectacular degree of glaciation – some of the finest cirques, hanging valleys and waterfalls to be found anywhere. Nowadays, few sizeable glaciers remain but the effects of past ice ages are clearly visible, and they still influence the present land-use and settlement pattern.

To reach the most flowery parts of the central Pyrenees you must leave behind the warm, settled lowlands, and climb through steep unglaciated and usually forested V-shaped middle valleys until you reach the higher elevations. Here, there are fewer villages, a largely pastoral way of farming, and a lovely mixture of flowery meadows, pastures, wetlands and woods. Above the highest permanent settlements, the Pyrenees National Park begins, protecting largely uninhabited countryside, and around this lies a huge buffer zone, with 40,000 inhabitants. Most of the flowery enclosed hay meadows and pastures lie outside the fully protected area, though the buffer zone does provide some protection by looser agreements between farmers and the park authorities.

The best hay meadows are astonishing, not just for their colourful abundance of widespread meadow species such as Ox-eye Daisy, clovers, and Bird's Foot Trefoil, but for the additional presence of so many special plants like the lovely blue-purple Horned Pansy, the endemic large-flowered Gouan's Buttercup, Elder-flowered Orchids in all their glorious colour forms, columbines, rampions, and even Pyrenean Fritillaries. Wetter patches may have masses of Alpine Marsh Orchids, various spotted orchids, louseworts, and Grass of Parnassus – a gorgeous mixture. The best meadows are usually to be found at about 1,100–1,200 m such as on the Plateau de Saugués, south-west of Gèdre, or around Gavarnie.

Above the enclosed land, the high pastures begin, grazed by flocks of sheep or cattle, and lying mostly within the national park. Depending on grazing levels and

the winter weather, many of these grasslands are wonderfully flowery, with an enormous range of species. There may be hillsides of the lovely deep blue English Iris, drifts of china-blue *Hyacinthus amethystinus*, patches of dark Dusky Cranesbill, pale yellow Wolfsbane, Masterwort, yellow Oxlips and Cowslips, and a range of orchids including Vanilla and Burnt Orchids. Where the soil is shallow and rocky, or on cliffs, a quite different range of flowers includes mats of the lovely pink endemic Tufted Soapwort, the enormous white spikes of Pyrenean Saxifrage, several species of primula, and, in shadier places, the beautiful purple or violet Pyrenean Ramonda, a relict of past warmer climates that has survived here through the ice ages. On wetter cliffs, you can find great masses of both the violet-flowered Large-flowered Butterwort and the pale blue Long-leaved Butterwort, supplementing their meagre ration of nutrients by catching and digesting insects through their leaves. Higher still are tundra and dwarf alpine plants, including many endemics.

This is wonderful countryside to visit, full of flowers and wildlife, with many well-marked paths, and those fabulous towering ice-sculpted peaks always in view.

Opposite: A gorgeous dense stand of Yellow Turk's Cap Lily in a Pyrenean meadow.

Below: A typical high Pyrenean hay meadow, full of wonderful flowers, with the peaks of Gavarnie beyond.

Ordesa and Monte Perdido, Spanish Pyrenees

INFORMATION

Location | In the north of Aragon province, between the villages of Torla and Bielsa, running up to the French border. In part, borders the Pyrenees National Park, France.

Reasons to go | Lovely high pastures in exceptional mountain scenery; beautiful montane woodlands; a rich fauna of insects, mammals and birds.

Timing | Best between mid-June and mid-July, but of interest between April and September depending on altitude, aspect and habitat.

Protected status | The whole area falls within the Ordesa y Monte Perdido National Park, which is also a Biosphere Reserve and EU Special Protection Area.

Opposite top: The high limestone slopes above the Circo de Soaso are covered with a lovely mixture of alpines, including the striking pink form of rock-rose.

Opposite bottom: An elegant spike of Pyrenean Saxifrage jutting out from a high limestone cliff.

I once spent a couple of nights at the Goriz Refuge, perched dizzyingly on the slopes of Monte Perdido at about 2,200 m. To reach here, you have to walk the whole length of the stupendous Ordesa valley, followed by a steep climb up through incomparable scenery to the highest pastures. Spending a night at the refuge allows you to be in this wonderful high, remote, unspoilt area in the early morning, when mists fill the valleys below and eagles and vultures begin to circle up on the thermals.

Ordesa Park (officially the Parque Nacional de Ordesa y Monte Perdido) covers 156 km² of staggeringly spectacular mountainous limestone scenery, reaching up to 3,355 m at Monte Perdido, close to the French border. It's a heavily glaciated landscape, dramatically eroded into cirques and deep U-shaped gorges, all exhibiting the typical karst features of caves and potholes.

There is only one way to see the most flowery parts of Ordesa and that is by walking. The normal route is from the end of the road above Torla, from the café at Ordesa. From here, you walk steadily upwards through beautiful forests of pine, beech and Silver Fir with constant views of sheer limestone cliffs towering above you. Some lovely flowers grow here, especially in the clearings, but the major show is yet to come. In a few places there are wonderful displays of the endemic Pyrenean Ramonda, which may cover a boulder or cliff face with its gorgeous purple flowers, and plenty of shade-tolerant orchids, Solomon's seals and bittercresses to keep you going. Everything begins to change as you enter the heavily glaciated upper valley, at about 1,500 m; the forest opens out, and the higher peaks come into view for the first time. From here, you are in high pastures which come alive with a dazzling selection of flowers from late May onwards, tall flowery grassland becoming gradually shorter as you gain elevation. In the lower parts, there are purple columbines and rampions, beautiful pink and red-flowered wild roses, masses of the white spikes of St. Bernard's Lily and St. Bruno's Lily, Pyrenean Avens, lovely frothy pink flowers of Large-flowered Meadow-Rue, pale yellow Wolfsbane, clumps of dark red-flowered Wild Chives, Small Yellow Foxglove, and a good range of orchids. Damper places may harbour dark red marsh orchids, yellow Globe Flowers, the deep blue-black spikes of Alpine Bartsia and several species of red or yellow lousewort, or the pretty blue endemic speedwell *Veronica ponae*. In drier places, you can find mats of the stunning little lilac-blue Creeping Globularia and some of its larger relatives, combining prettily with yellow rock roses, saxifrages and pinks.

On the way up through the valley, you pass several steps in the valley floor, evidence of past glacial action, now home to beautiful little cascades, but none indicate any great change in the vegetation. However, when you reach the much larger Circo de Soaso, the change is quite dramatic. The limestone slopes around it are steep, and covered with rock roses, particularly the beautiful pink form of Common Rock Rose, with Mountain Avens, Ashy Cranesbill, several saxifrages, and Pink Sandwort. The steep cliffs support impressive populations of the striking Pyrenean Saxifrage, with lovely white drooping flower spikes up to 60 cm long.

If you have the time and energy to climb on up beyond the cirque, you eventually reach the high pastures and tundra that soon lead to the refuge. This is the place for mats of Purple Saxifrage, intense blue Spring Gentian, purple Alpine Snowbell, cushions of the pretty pinkish Pyrenean Whitlow-grass, and occasionally the gorgeous yellow, orange or red flowers of Pyrenean Poppy. Some of these, with other high altitude specialists, continue on to the very highest levels, following the melting snow upwards.

48 The Picos de Europa

INFORMATION

Location | Just inland from the north coast of Spain, the Costa Verde, south-west of Santander, and due south of Llanes.

Reasons to go | Spectacular glaciated high limestone mountain scenery with abundant flowers, especially in the hay meadows. Rich and varied bird and mammal life, unspoilt villages and traditional way of life.

Timing | Remains flowery from spring to autumn, best displays in late May and through June.

Protected status | Much of the area is now a national park and Biosphere Reserve, though not yet fully protected.

It is said that the name Picos de Europa (Peaks of Europe) derives from medieval times when land-hungry Basque fishermen returning home from long periods at sea had their first view of land. Nowadays, many visitors have their first glimpse of these snowy high peaks from the ferry entering Santander, but the effect is the same: they exert a strong magnetism, drawing you in for a closer exploration.

Right: The autumn-flowering *Crocus serotinus ssp. asturicus* growing in masses in a high altitude pasture.

Opposite: A classic Picos high meadow, with masses of Asturian Yellow Rattle and other flowers below the high limestone peaks.

Opposite: The striking spikes of Heart-flowered Serapias orchid are more abundant in the Picos meadows than anywhere else.

Although most of northern Spain is mountainous, the Picos stand out as something special. The peaks are mainly sharply eroded limestone, in contrast to the more rounded shales and schists of the surrounding mountains, and they are higher, reaching 2,648 m at Torre de Cerredo, with many other peaks rising above 2,500 m. Their great height, and their position on the Atlantic coast, means that they intercept most of the depressions that sweep in from the Atlantic; these are notably wet mountains and it may snow on 100 days per year at the higher levels. Although this high level of precipitation may mean that a visit is likely to be affected by rain, it has the advantage that the mountains remain wonderfully green and flowery for most of the summer, and snowy for most of the winter. It has probably also contributed to the survival of a largely traditional way of rural life, because intensive farming in this climate and topography is very difficult.

The hay meadows are probably the most strikingly flowery feature of the Picos; recent studies have recorded over 600 species of flower in the meadows, and they are considered to be one of the most intensely flowery temperate habitats in the world. The best hay meadows, such as those around Espinama and Fuente De in the Deva valley are simply glorious – an ever-changing tapestry of reds, blues, yellows and white. Around 50 species of orchid are found here, including some hay meadow specialists such as the deep reddish-brown Heart-flowered Serapias – often in extraordinary numbers – Pink Butterfly Orchid, Burnt Orchid and Green-winged Orchid. Other typical hay meadow flowers include Greater Yellow Rattle (in its Asturian subspecies), pale yellow Kidney Vetch, tall spikes of White Asphodel, the lovely reddish flowers of Bloody Cranesbill, Cowslip, deep blue heads of Round-headed Rampion, Winged Broom and dozens of others. In damper areas, there may be several species of marsh orchid, beautiful pale yellow globe flowers, several deep purple thistles, Marsh Marigolds and Whorled Lousewort. Where the meadows are shadier, you may find drifts of the pretty china-blue Pyrenean Squill, Herb Paris, Bastard Balm and various lungworts. On more acid soils, you are more likely to find Maiden Pinks, St. Dabeoc's Heath, Kerry Lily, Sheep's Bit and Meadow Saxifrage.

The high pastures and rocky outcrops of the Picos are quite different, less lush but with many special species. There are few high road passes in the Picos, and the best way to get high is via the stupendous cable car from Fuente De, giving access to beautiful snowy glaciated high mountain country. This is the place for tiny dwarf Narcissi, intense blue Spring Gentians, deeper blue trumpet gentians, rock jasmines, Monte Baldo Anemone and rather similar pasque flowers, and clumps of lovely saxifrages. The flowers are quite good enough on their own, but the Picos are also spectacularly beautiful, fascinatingly historic, and full of birds, butterflies, mammals and many other creatures.

52

Sierra de Grazalema, Andalucia

Opposite top: The pretty Three-coloured Bindweed is often abundant in grasslands and roadsides throughout these mountains.

Opposite bottom: Some of the olive groves here can be astonishingly flowery, such as this one, dominated by masses of beautiful Mediterranean Catchfly.

There's something rather special about the Sierra de Grazalema. It may be the beautiful old white villages appearing to grow from the limestone cliffs and rocks; or the luxurious vegetation in an otherwise dry area, thanks to the unusually high rainfall; or the lovely displays of flowers, the unspoilt forests and cork oak groves, and the abundance of birds. Whatever it is, this relatively low mountain range exerts a strong magnetism on anyone who goes there – and almost everyone returns for another visit.

The park covers the core of the mountains, though surrounding areas, especially to the north and east, are also of interest. The park is also contiguous southwards with the huge Alcornocales Natural Park, which runs right down to the coast at Spain's southernmost point. Though not especially high, the Grazalema mountains are the first significant peaks in the line of depressions coming in from the Atlantic, making them unusually wet – in fact, one of the wettest parts of Spain, with an average rainfall of up to 2,000 mm. The flora is notably rich, with at least 1,400 species in the park alone, of which at least a dozen are confined to this area, and many others are endemic to southern Spain and north Africa.

Thanks to its southerly position, the shows of flowers begin early in the year. The beautiful blue *Iris planifolia* can be abundant through January and February, followed soon afterwards by some gorgeous masses of yellow and white narcissi. Many shady limestone cliffs have sheets of yellow dwarf daffodils in March, growing alongside Paper-white Narcissus, the earliest orchids and pretty brightly coloured Yellow Anemones. An interesting plant that looks its best through late winter and early spring is the Red-berried Mistletoe, like Common Mistletoe but with bright red berries, parasitic on hawthorns and olives.

From about mid-April through much of May some of the displays can be quite spectacular. The mixture and amount of flowers vary according to the winter rain and the level of grazing, but there is always somewhere of interest at this time of year. Damper cliffs and rocks are likely to be vibrant with white spikes of the tall endemic star-of-Bethlehem *Ornithogalum reverchonii*, lovely blue Spanish Bluebells, several white saxifrages, the attractive Intermediate Periwinkle, a large pink-flowered valerian, Kidney Vetches, beautiful two-toned Wild Peas, and lots of orchids. About 30 species of orchid grow in the park, and April is definitely the best month to see them, as Lange's, Champagne, Italian, Bee, Man, Pink Butterfly and several tongue orchids add their distinctive shape and colour to the mix. In early May, you can see Spanish Iris, buglosses – of which the lovely silver-leaved, pink and blue-flowered *Echium albicans* is pre-eminent – beautiful pinkish-purple Western Iberian Paeonies, Purple Phlomis, Large Blue Alkanet, and sheets of yellow broom. At about the same time, some of the drier limestone cliffs burst into colour as the endemic pale yellow Shrubby Buckler-mustard comes into flower in extraordinary quantities, often in company with a purple woundwort *Stachys circinata* and a distinctive shrubby yellow-flowered endemic knapweed.

An important component of the Grazalema conservation tapestry is the remnant forest of Spanish Fir, high on the north slopes of Sierra del Pinar. A walk up to the forest (for which a permit is needed) takes you through a range of flowers including sometimes-spectacular displays of the prickly blue Hedgehog Broom, masses of the endemic orange-red poppy *Papaver rupifragum*, another species of paeony, and good quantities of Spanish Fritillary.

Cape St Vincent area

INFORMATION

Location | The far south-western corner of Portugal, either side of Cape St Vincent.

Reasons to go | Beautiful rugged coastal scenery, unspoilt by south-west European standards, with a rich flora including many endemics and fine displays.

Timing | Variable, but generally good through March and April, with the best displays in April.

Protected status | The whole area lies within Parque Natural do Sudoeste Alentejano e Costa Vicentina, though this gives rather limited protection.

At the extreme south-western margin of Europe, where the hard rocks of the Algarve meet the storms of the Atlantic, there is a lovely wild unspoilt area, alive with flowers amongst spectacular coastal scenery. This is the Cape St Vincent area, part of the much larger Alentejo and Costa Vicentina Natural Park which extends for about 40 km northwards and 20 km eastwards from the Cape, and encompasses some of the finest coastal scenery in Iberia.

The gem of the area for flower displays is a high level plateau of hard dolomite, edged with sheer 150 m cliffs and topped with a version of limestone pavement. For at least half of the year, it can look featureless and barren, but as the winter rains subside and the warm southern sun strengthens, a wonderful variety of flowers bursts from the cracks to give colour and form to the landscape. Amongst the first, in February and March, are the daffodils, particularly beautiful golden clumps of Hoop Petticoat Daffodil in the form known as *Narcissus obesus* reaching right to the edge of the cliffs. These are joined or followed by Spanish Bluebell, white garlic, pale blue spikes of St Vincent Squill, pinky-purple stocks, the rare hyacinth-like *Bellevalia hackellii*, magenta-pink Shore Campion, orchids, especially the gorgeous Mirror Orchid, and many other herbs and bulbs.

In this windswept landscape, woody plants are reduced to low mounds, which produce a wonderful intensity of flowers. Many species are so covered with

Right: A gorgeous mixture of spring flowers, mainly Italian Orchids and *Halimium commutatum*, on the cliffs near Burgau.

Opposite: Wind-pruned spring flowers, especially Small-flowered Gorse and Rosemary, on the clifftop at Cape St. Vincent.

flowers in spring that it is hard to see the foliage at all. Intense blue cushions of Rosemary intermix with the devastatingly spiny cushions of Tragacanth, covered with sweet-scented white pea flowers, or tight spiny cushions of the yellow-flowered Small-flowered Gorse. One of the commonest plants is an endemic shrubby white-flowered cistus, with dark glossy sticky aromatic leaves, most frequently known as *Cistus palinhae*, though also classified as a subspecies of Gum Cistus. A close relative, *Halimium commutatum* grows everywhere as low hummocks tightly covered with beautiful bright yellow flowers. Scented yellow sprawls of *Coronilla valentina*, a common garden plant native here, can be seen from miles away, and throughout, the tall red flower spikes of Wild Snapdragon emerge from the bushes, after using them for support and protection. Other endemics include the tiny but delightful flower-covered cushions of Diamond Flower or Violet Cress, growing along tracks or in winter-wet depressions, and two endemic brassicas.

Northwards from St Vincent are many beautiful areas of coastal habitat, rich in flowers, especially around Carrapateira and Aljezur where lovely dunes and grasslands are alive with grape hyacinths, Three-leaved Snowflake, and many other flowers. Eastwards along the south coast, the effects of development are greater, and the habitats are more fragmented, but beautiful areas remain, such as around Boca do Rio and Burgau, where orchids, halimiums and narcissi run riot along the cliffs.

Opposite: A striking clump of an early-flowering narcissus, *Narcissus obesus* on the Cape St Vincent cliffs.

Below: Even in early march, the bare clifftop grasslands become a vibrant mass of colour.

58 The Upper Engadin Valley

INFORMATION

Location | South-east Switzerland, around St Moritz and Pontresina.

Reasons to go | Spectacular displays of flowers in hay meadows, alpine grasslands and screes, and on the high cliffs and snow-melt areas.

Timing | Good from April until October, but best from late June through July.

Protected status | Patchily protected but generally not threatened. The only Swiss national park lies nearby, though it is less spectacularly flowery than the area described.

Opposite: Intensely flowery mats of pink-flowered Creeping Azalea growing with pansies and pasque flowers on the Albula Pass.

Few places are so closely associated with beautiful alpine flowers as the Swiss Alps. There are dozens of flowery areas in the mountains of Switzerland but for sheer variety, abundance and spectacle, nowhere in the country quite matches the Upper Engadin in the far east. Here, it seems, almost everywhere is flowery, from the lush valley floor and lower-slope hay meadows to the highest peaks, which reach to over 4,000 m. The climate is good, with some parts claiming 320 sunny days per year, the highest number in Switzerland, though the definition of 'sunny' may be flexible. The infrastructure for visiting the mountains and other sites is outstanding, with an excellent coordinated system of buses, trains, cable cars and other means of public transport. Roads through the valley and over several mountain passes make it very easy to reach the altitude most appropriate to the time of year you are visiting, and in high summer it is possible to be transported to heights of over 3,000 m – a breathtaking experience in every way.

Many of the lower valley hay meadows are outstandingly beautiful for their flowers, a fabulous tapestry of pink, yellow, blue and orange, with masses of sainfoin, yellow rattles, rampions, orchids, campions, clovers, pink Bistort, Viper's Bugloss, Ox-eye Daisy, bellflowers and even occasional Orange Lilies.

Above the meadows, there is often a belt of woodland dominated by larch, and colourful with pink daphnes, wintergreens and orchids, though the real mountain show starts higher. Whether you are on granite, limestone or other rock these high slopes all burst into flower as the winter snow recedes, producing sheets of colour. Two species of snowbell grow in abundance often pushing up through the snow before it has melted, and at least four species of pink to purple primulas grow in drifts with yellow Oxlips and Bear's Ear Primulas, wonderful quantities of Spring Pasque Flower and the taller Alpine Pasque Flower, frequently in its lovely lemon-yellow form. Amongst them may be clumps of deep blue Trumpet Gentians or intense blue Spring Gentians, mats of Mountain Avens covered with large white flowers, and patches of sprawling yellow cinquefoils. On more acid soils, such as granite, there may be extraordinary quantities of Creeping Azalea. Though its pink flowers are only small, it more than makes up for this by producing thousands of them, on mats no more than a few centimetres high, frequently intermixed with a beautiful mixture of arctic-alpine lichens.

A few flowers specialize in growing at really high altitudes, only starting at heights of over 2,000 m, often continuing up to the limits of plant growth at over 4,000 m. These are usually strikingly beautiful plants; they grow in tight cushions

to escape the worst ravages of the severe climate, then produce their flowers in intense patches of colour for the short summer season. On high cliffs and rocky slopes, bright blue patches of King-of-the-Alps (a relative of forget-me-nots) grow with pink Moss Campion, or occasional cushions of one of the gorgeous dwarf pink or white rock jasmines. On loose scree and around receding snow patches, there may be carpets of the lovely white-flowered Glacier Crowfoot, its flowers turning pink as they age, growing with white or purple saxifrages, and the bright orange cups of Creeping Avens. This is an extraordinary and exhilarating floral landscape at the outer limits of plant growth, well worth the effort of the climb.

Opposite: Masses of the elegant Glacier Crowfoot flowering just after the snow melts at over 3000 metres.

Below: Mid-altitude hay meadows in the Engadin are some of the most flowery to be found anywhere, with rampions, sainfoin, yellow rattle and many other flowers.

The Dolomites

Location | North-east Italy, beginning about 70 km north of Venice and running up to the Austrian border.

Reasons to go | Exceptionally beautiful and spectacular mountain scenery, with fabulous displays of flowers almost throughout.

Timing | Early April to October, but best from early June to late July.

Protected status | There is one national park (Dolomiti Bellunesi) and several natural parks and nature reserves in the area, giving a reasonable degree of protection; 1,420 km² are designated as a World Heritage Site.

The Dolomites must be one of the most spectacular and attractive mountain ranges in the world. They may not be particularly high, but they have a sublime form which overwhelms and humbles any visitor. One of the main reasons for their distinctive shape is their geology – they are composed of the uplifted and eroded remnants of a huge block of rock known as dolomite (the mountains are named after the rock rather than vice versa).

The distinctive feature of dolomite is the high level of magnesium, in addition to the calcium normally found in limestone. The presence of the rock has a marked effect on the flora and makes an interesting comparison with equivalent limestone, which hosts a different combination of plants. Some species flourish on dolomite, others cannot tolerate it, and some acid-loving species, such as Least Primula, can occur on dolomite but not on limestone. Dolomite also gives rise to cliff- and rock-faces that are particularly suitable for plant growth, and many are botanically rich.

The flora comprises about 2,000 species (depending on how you define the area). The number of endemics is surprisingly low, probably because the Dolomites are neither isolated geographically from other mountains, nor geologically, since dolomitic rocks occur elsewhere in the region. There are several exceptionally flowery habitats here, particularly the mid-altitude hay meadows, the high alpine pastures, open woodlands, and cliffs and rocks. Most of the hay meadows are still unspoilt, with a beautiful sward that includes Orange Lilies, several rampions, many orchids, Cowslips, bellflowers, louseworts, restharrows, meadow-rues, yellow rattles, and many others, depending on the degree of dampness. The meadows are generally best in the second half of June.

Higher up lie the unenclosed alpine pastures, covering a huge area, often interspersed with patches of woodland, large rocks, screes and scrub, and frequently ending abruptly at a sheer cliff. These places are stunningly colourful, with a whole gamut of plants ranging from most of the hay meadow species to high altitude specialists such as snowbells and rock jasmines. However, most are grazed by cattle or sheep, and it is hard to predict when they will be at their best – some time in late June and July will usually be about right. The rocks and screes within and around the pastures are often intensely flowery with gems like the gorgeous pink Dwarf Alpenrose, bright blue Alpine Forget-me-not, golden yellow Rhaetian Poppy, saxifrages, Bluish Paederota, and occasionally the emblematic and enigmatic Devil's Claw, related to rampions, but unlike any other flower.

In a few places, the older schists and igneous rocks are exposed, such as at the Rolle Pass, or the cliffs east of the Pordoi Pass. These can be spectacularly flowery too, but with different plants – golden Arnica, butterworts, trumpet gentians, yellow Creeping Avens and Glacier Crowfoot in the grasslands and screes, and beautiful cushion species on the cliffs such as the intense blue King-of-the-Alps or the delightful yellow Vitaliana.

There is little need for a list of localities in the Dolomites, as so much of the area is both spectacularly beautiful and intensely flower-filled. Above about 1,000 m everywhere is good – it's just a matter of timing.

Below: Clumps of the stunning pink Dwarf Alpenrose on the slopes of the Tre Cime de Lavaredo.

Overleaf: A gorgeous montane hay meadow on the Tre Croci pass above Cortina d'Ampezzo.

66   The Lake Garda mountains: Monte Baldo and Monte Tombea

Opposite: Mountain Valerian, Alpine Basil-thyme, one of the mountain cornflowers and many other flowers on limestone scree on Monte Baldo.

Monte Baldo, on the eastern shore of Lake Garda, is a spectacular little gem of a mountain, whale-backed in form but eroded into sheer cliffs, deep valleys and stark peaks. A rather complex mixture of limestones, dolomite, and other rocks gives rise to a lovely mixture of largely unspoilt habitats – subalpine grasslands, scree, cliffs and other rock outcrops, pastures, hay meadows and mixed woodlands, all rich in flowers.

The wonderful bonus here is that so many of the flowers are particularly striking or unusual: dozens of orchids, two lovely paeonies, narcissi, special rock plants such as the extraordinary Devil's Claw in some abundance, the attractive blue speedwell-relative *Paederota bonarota*, a rare and beautiful daphne *Daphne petraea*, the Monte Baldo Anemone, several lilies and hundreds of others, in a dazzling mixture. Running down the west side of the lake is the almost equally interesting, but less well-known, Monte Tombea.

A high proportion of the most attractive plants are found on the high screes and cliffs, some widespread above about 1,500 m, others confined to particular altitudes. The fabulous pink-flowered silver-leaved cushions of the cinquefoil *Potentilla nitida*, are often found growing together with one of the most attractive members of the heather family, the gorgeous pink or red-flowered Dwarf Alpenrose, a plant of high rocky limestone places in the eastern Alps. The intriguing Devil's Claw – actually a distinctive relative of the rampions – is common on cliffs here, producing its curiously beautiful clusters of bluish-pink, black-tipped flowers in July and August. Other special plants include two near-endemics, the lovely pink Rock Daphne, and the white-flowered *Callianthemum kernerianum*, as well as the rare and beautiful pink Large-flowered Campion. A lovely waymarked walk, the *Sentiero delle Cime del Ventrar* begins at 1,564 m near the top of the cable cars, specifically designed to allow access to some of the cliffs. It's not a difficult route, though it does demand a head for heights, as there are some dizzying drops along the way.

Vast areas of grasslands on these mountains vary from alpine pastures with Pheasant's Eye Narcissus, St Bruno's Lilies, Yellow Gentians, Globe Flowers, bluish-purple Alpine Asters, clear red Alpine Roses, Trumpet Gentians and Globe-flowered Orchids (all often in abundance) to lusher lowland pastures and meadows, full of orchids, Southern Greenweed, Martagon Lilies, perennial cornflowers, Clustered Bellflower, Round-headed Rampion, two species of columbine and a profusion of other species. In a few rocky sunny grasslands, masses of the lovely red paeony

Paeonia officinalis look wonderfully exotic. The edges of the woods are also rich, with masses of Bastard Balm, yellow foxgloves, another paeony, hellebores, orchids, fragrant beds of Lily-of-the-Valley and striking small trees of Alpine Laburnum, to name but a few.

Access is reasonably easy. For Baldo, there is a narrow surfaced road along most of the east side, with many paths and a cable car leading to higher areas. A long and popular cable car up from Malcesine on the shores of Lake Garda gives easy access to high parts, though the area around its top station is very busy at weekends and holidays. For Tombea, there are minor roads up from both sides, and the Passo de Tremalzo and the nearby old military road are especially good.

Opposite: A mass of Alpine Aster and Livelong Saxifrage with the steep west-facing cliffs of Monte Baldo beyond.

Below: A wonderful clump of red paeonies growing in the protection of a juniper bush on limestone scree.

70

Piano Grande, Monti Sibillini National Park

Location | The southern part of the Monti Sibillini National Park, east of Norcia, around the village of Castelluccio.

Reasons to go | Good general mountain flora, exceptional displays of cornfield weeds and grassland flowers in the Piano Grande. Wildflower festival in June.

Timing | Peak displays from mid-May to late June in the piano areas, with steady progression of mountain flowers through spring and summer.

Protected status | The area lies wholly within a National Park, and the protection level is high, with good management.

As you drive up through the wooded slopes of Monte Sibillini from the beautiful walled town of Norcia towards the hill town of Castelluccio, there is little hint of what awaits you at the top. Many similar mountain slopes in the Apennines of northern Italy are beautiful and flowery, though not exceptional. But as you crest the ridge at 1,500 m and have the first glimpses of the great bowl that makes up the Piano Grande, you realize immediately that this is different and special. Whether it is filled with dense mist, dominated by dark thunder clouds, or clear and sunny, that first view is astonishing.

Right: A fine clump of blue Hepatica flowering in early spring in the beechwoods above Piano Grande.

Opposite: A breathtaking display of cornfield weeds, especially Cornflowers, mayweeds and Field Poppy on the floor of the Piano Grande.

Above: Even the fields on the lower mountain slopes are full of flowers, with masses of Narrow-leaved Vetch on the bank beyond.

The Piano Grande (meaning great plain) is a classic example of a *polje*, or inwardly draining valley that has no river outlet. These occur in many south European mountains, especially in Greece, Italy, and countries of the former Yugoslavia, where the limestone rock is porous enough to allow the rain and snowmelt to sink into the ground rather than accumulate into an eroding river. In the cooler, wetter climates following the ice ages, the water formed into a lake, collecting more quickly than it drained; as the silt gradually built up, and the climate warmed, the amount of standing water decreased and the amount of dry land steadily increased. Not surprisingly, such areas are both flat and very fertile, and they frequently lie at an altitude that supports cultivation, so they have become a target for colonization and farming. The Piano Grande is a huge example of this type of plain, some 10 km across, surrounded by high limestone peaks on all sides.

Many such poljes have an interesting flora because their remoteness and harsh winter weather, coupled with occasional flooding, have prevented intensive arable farming from taking place. The Piano Grande is no exception, with its rich pastoral flora of Pheasant's Eye Narcissus, Elder-flowered Orchids and fritillaries; but it is exceptional because the arable flora of a traditional way of farming has been semi-fossilized by the creation of the Monte Sibillini National Park. The drier parts of the plain are under arable cultivation, in a series of small strips and squares, representing fragmented ownership and rights. Rarely do two adjoining strips support the same crops (though the majority of them are the rather special local variety of lentil), so each is under a different regime of ploughing and management. In the absence of pesticides, the cropped strips support a marvellous flora of annual 'weeds' and short-lived perennial species, and the patchwork of crops gives rise to an astonishing corresponding patchwork of colour. One field may be dominated by the vivid blue of Cornflowers, perhaps with scarlet swathes of Field Poppies on either side, and a wonderful honey-scented mass of pink Sainfoin nearby. A little later, the purple of Corn Cockle may come to dominate, perhaps mixed with White Campion and the yellow spikes of Common Toadflax. The headlands separating some of the fields, slightly raised up by centuries of ploughing, are often covered with a mass of blue-purple Narrow-leaved Vetch punctuated by spikes of Orange Mullein. Although all these species are quite common in southern Europe, nowhere else do they occur in such reliable quantity, and in such a wonderful mosaic of colour.

Many less common plants can be found in the fields, from the tiny rosettes of Annual Androsace to the more obvious delights of Large Venus' Looking-glass and some of the rarer poppies. The uncultivated hills around the plain also have a fine montane flora; in spring, crocuses, orchids, Cowslips and Yellow Star-of-Bethlehem mass in the grasslands, anemones and hepaticas form drifts in the woods, followed later by a rich mix of mountain plants. There are a number of other 'pianos' in the area though none is quite as spectacular as Piano Grande.

The village of Castelluccio holds a flower festival in the middle of June, usually coinciding with the best show. May is perfect for the narcissi, fritillaries and earlier orchids.

Below: Some of the roadsides on Piano Grande are astonishingly flowery, with an explosion of colourful annuals in summer.

74

The Gargano Peninsula

INFORMATION

Location | A large oval peninsula projecting east from Italy into the Adriatic Sea, just north-east of Foggia.

Reasons to go | Marvellous displays of spring flowers, extending into early summer, in a lovely ancient hilly landscape. Particularly good for orchids and irises.

Timing | Peak displays from early April to early May, though autumn is good for a few species.

Protected status | Although nominally protected as the Gargano National Park, action and protection on the ground is limited and some species are still declining.

Opposite: A typical dazzling Gargano field, full of irises, orchids and other colourful flowers in April.

You approach Gargano almost as if it were an island, crossing the great plains around Foggia until its huge bulk rises above you, often shrouded in mist. And when you do arrive, it feels like an island, geographically, culturally and ecologically detached. It is an astonishing place, full of beauty and history, where the record of thousands of years of farming can be read in the pattern of small fields, ancient stone walls and historic buildings.

Over 2,000 species of flower grow here, an impressive total for a small limestone peninsula, but more striking is their sheer abundant exuberance. You feel that plants want to grow here, whatever the conditions. Some places reveal their treasures only slowly, and only after hard work: not Gargano – they assault you from every angle. The area is probably best known for its orchids, which occur in a dazzling and confusing variety and quantity; over 80 species of orchid are said to grow here (though much depends on what classification system you follow), and they are everywhere, especially in April and early May. Widespread species such as Pink Butterfly, Green-winged and Man Orchid overflow from the fields, interspersed with greater or lesser amounts of some of the special orchids of Gargano – the eponymous *Ophrys garganica*, the much rarer dark reddish brown *O. sipontensis* (named after the little village of Siponto), or the lovely *O. apulica*, all enigmatic spider orchids. For the orchidophile, Gargano is a source of endless delight, but it is a pleasure for the non-specialist, too.

Gargano is almost entirely hard limestone, closer in origin to the limestones of Serbia and Montenegro across the Adriatic Sea than those of neighbouring central Italy. Small stone-walled fields and unfenced ancient common pasture are the most prominent habitats. The flower show depends on the weather and the degree of grazing, but a good year will be astonishing in colour and variety. Some fields may be a mass of purple, blue and yellow irises, dotted with orchids and interspersed with the red and white of the semi-parasitic Southern Red Bartsia, or a froth of yellow buttercups. Others may be dominated by orchids, together with drifts of lovely white spikes of Meadow Saxifrage and patches of the striking yellow and black vetch *Vicia melanops*.

Higher parts of the peninsula are dominated by largely deciduous ancient woodlands. The best of these, such as Bosco de Quarto or Foresta Umbria, have a wonderful spring flora of carpets of blue or white Apennine Anemone, scented Pheasant's Eye Narcissus, hepaticas in many colours, Yellow Marsh Orchids, wild garlics (especially *Allium pendulinum*), beautiful red paeonies, and many others,

while in late summer there are masses of pinkish-purple cyclamen. Cornfields are still often full of colourful weeds such as Cornflowers and poppies, and some of the damper ones are dotted with the increasingly rare yellow Wild Tulip. Even the cliffs and old shady walls, such as the north-facing outcrops and castle walls at Mont St Angelo, are a mass of flowers: here there are huge patches of blue-purple aubretia, yellow Leopard's Bane, Dense-flowered Orchids, and the lovely Gargano Dead-nettle growing in great pink clumps.

Many contributing factors explain why Gargano is so good. The hard, rather infertile, limestone favours the growth of less competitive plants in place of more vigorous grasses, and has also restricted agriculture. The long history of settlement has created a lovely variety of habitats but has rarely been intensive enough to destroy the rich diversity of plants that has arisen. The climate plays a significant role, too; warm enough to favour a big range of species, yet damper than many Mediterranean areas. Whatever the reasons, the result is wonderful, one of the most alluring wildflower destinations in Europe.

Opposite: Old common land with a wonderful display of Pink Butterfly Orchids, with scattered dwarf irises, Hyoseris and many other flowers.

Below: A Green-underside Blue butterfly perched on a Green-winged Orchid.

Bottom: Spring in Gargano: four species of orchids, an iris and other flowers.

The Julian Alps

Opposite top: The lovely nodding flowers of Alpine Clematis sprawling over a rock.

Opposite bottom: Ox-eye Daisies and Viper's Bugloss fill an old hay meadow by Lake Bohinj.

At the point where Slovenia meets Italy and Austria, towards the eastern end of the great alpine block, there is a lovely area of high limestone mountains known as the Julian Alps. This beautiful area is an intriguing blend of western, eastern and Mediterranean cultures, with a rich history and farming heritage. Hay is dried on old wooden hay-racks, and every house has its vegetable garden, fruit trees and stacks of wood for the winter. Until recently, it had also escaped the ravages of mass tourism that afflict the most popular parts of the Alps, though this is changing, particularly since Slovenia became part of the EU.

The flowers have survived thanks to the protection of the national park and the continuation of traditional farming, encouraged and supported within the national park buffer zone. The flora is very rich, with a wonderful mixture of general alpine flowers, eastern alpine flowers, widespread lowland species, and an element of Mediterranean influence thanks to the warm summers. There are a few uniquely Slovenian endemics, such as Triglav Hawksbeard and Julian Poppy, but most of the endemics also encompass a wider area of the eastern Alps.

The first flowery habitat you reach, coming from any direction, is the hay meadows, particularly above an altitude of about 500 m. Some have been fertilized and improved into a monoculture, but the vast majority remain beautifully colourful whilst continuing to provide good quality hay. Viper's Bugloss forms intense blue swards in many of the meadows, often growing with purplish-blue Meadow Clary, yellow rattles, Hoary Plantain, Swallow-wort, purple wild thymes, yellow rock roses, the dark flowers of Dusky Cranesbill, blue Clustered Bellflower, several species of blue speedwell, and a dozen or more species of orchid. In shadier or damper places, you may find the fabulous orange spikes of Carnic Lily, with fragrant Lily-of-the-Valley, marsh orchids, Marsh Valerian and German Asphodel.

Above the hay meadows are extensive forests of beech and Norway Spruce – a wonderful habitat rich in wildlife, but rarely spectacularly flowery. Continuing upwards, you eventually reach the higher mountain pastures, such as those on the Vrsic Pass, at the top of the cable car on Vogel, or on any number of high walks towards Triglav. Such places can be full of treasures: Dark Columbine, pale yellow Globe Flowers, delicate pale yellow Oxlips, Common and Pyramidal Bugle and Cypress Spurge jostle with Common Spotted Orchids, Fragrant Orchids, Water Avens, Drooping Bittercress and the lovely curving spikes of Lesser Solomon's Seal. Early in the year, crocuses and the clear white (or occasionally pink) flowers of

Christmas Rose appear as the snow melts, and autumn brings masses of Meadow Saffron. Higher still are more alpine specialists such as Dwarf Alpenrose, flower-covered cushions of the pink-flowered *Potentilla nitida*, slaty-blue globularias and many bellflowers and saxifrages.

In the last few years, the beautifully flowery nature of this area has been recognized by the establishment of the Bohinj International Wildflower Festival, held each year towards the end of May. It has steadily grown in size and scope, and now offers a varied programme of guided walks, workshops, seminars, concerts and other flower-related events.

The grasslands of southern Transylvania

INFORMATION

Location | South of the city of Sighisoara in an area roughly bounded by Sibiu, Fagaras, and Rupea.

Reasons to go | Exceptional displays of flowers, and abundant insects, in a beautiful, historic pastoral landscape.

Timing | Of interest almost any time between April and October, but most colourful in June and early July.

Protected status | Most of the area is in a Natura 2000 site, though how successful the protection afforded by this designation will be is not yet clear. Survival of the farming pattern within the modern rural economy is crucial to maintaining the landscape and its biodiversity.

Opposite: Black Broom and other flowers clothe the curious 'tumps' near the village of Apold.

To visit the old Saxon villages of Transylvania is to be back in medieval rural Europe, in a time when each settlement depended on the countryside around it. Nowhere else in Europe is it possible to so clearly understand the medieval farming system and to see the plants and animals which depend on it.

The story of Saxon Transylvania is a strange one; in the 12th century, King Geza of Hungary began to invite Saxons over to the area that is now Transylvanian Romania, with the idea that they would assist in defending the Hungarian Empire against attacks from the Tartars, as well as providing much-needed agricultural and mercantile expertise. He provided land, and over the next few centuries, they steadily occupied this area, building both cities and villages. The rural villages were tight-knit, well-organized settlements, surrounded by arable land close to the village, hay meadows in fertile areas, vast communal grazing pastures, and woodland on the hilltops, all used sustainably by the villagers. This system was prevalent elsewhere, but it is only really in Romania that it has survived, and even the villages themselves are of a type no longer found elsewhere in Europe. The agricultural system was disciplined and efficient, but because it was based on extensive grazing of cattle and sheep, and there were no modern fertilizers or pesticides, the resulting grasslands were astonishingly flowery – and they still are.

This is almost certainly the largest area of flower-rich lowland grasslands left in Europe. Stand, for example, on a hilltop above the classic village of Viscri, in June, and you can see almost nothing but unspoilt pastures, broken by occasional woodlands, extending away towards the Carpathian mountains on the horizon. It's a fascinating place socially and historically, and also a wonderful place for the naturalist, with exceptional numbers of flowers, including rare and threatened species, and many butterflies and other insects. Some experts believe that the best of these grasslands have as high a density of flowers as anywhere in the world. Whatever the figures, it's a lovely place to be in at almost any time between spring and autumn.

At its flowering peak, the grasslands may be yellow with hay rattles or Lady's Bedstraw, pink with Sainfoin or Crown Vetch, purple with Military Orchids or Great Milkwort, or blue with Meadow Clary and Narrow-leaved Vetch. Some places may be carpeted with yellow Dyer's Greenweed or Winged Broom, creamy masses of Dropwort, or the blue or yellow of flaxes. In places that are steeper or drier, many rare species occur, sometimes in surprising abundance – Red Viper's Bugloss, Burning Bush, Entire-leaved Clematis, Yellow Pheasant's Eye, several species of

cow-wheat, the bluebell-blue of Nodding Sage, bellflowers, and an abundance of orchids, to name but a few. By late June or early July, butterflies abound and it's also a good place to see birds.

Despite its wonderful richness, and air of bucolic stability, the area is under great threat, as many Saxons have returned home to a reunified Germany, and EU regulations are forcing unwelcome changes on the farmers. The whole area of interest has recently been declared a Natura 2000 site, which provides some hope for its future, and many organizations are working to protect it. Let's hope they are successful, for this is a really special area.

Opposite: Butterflies, such as these Silver-studded Blues, are so abundant here that they need multi-storey roosts.

Below: An astonishingly flowery pasture dominated by Great Milkwort and Dropwort in eastern Transylvania.

84 Mount Parnassus and Delphi

Opposite: The glorious view south from Delphi to the Gulf of Corinth, filled with Jerusalem Sage and other spring flowers.

Just the names Mount Parnassus and Delphi are enough to set the heart racing, with their close associations with the ancient world of Greek myth and legend. What's more, Parnassus has been a national park since 1938, and is undoubtedly one of the finest botanical sites in Europe, with a huge range of species (including many Greek endemics, and ten plants found only around Parnassus) and some fabulous displays on the lower slopes.

However, the park was declared primarily for cultural reasons and excludes the highest and many of the botanically richest areas. It is very small, covering only 3,500 ha, and there is virtually no management, interpretation or real protection. There are two ski centres on the mountain, development is steadily creeping across the high valley of Livadia towards the Park edge, and there is a bauxite mine on the northern slopes. Although once forested, the forest is now fragmented and subject to continuing logging. The importance and beauty of the mountain are under severe and continuing threat but despite these problems, it is still a fabulously flowery place, and one bonus of the ski centres is the easy access to the high places, which is relatively rare in Greece.

Parnassus rises high above the Gulf of Corinth, giving magnificent wide views across to the Peloponnese and northwards into Epirus. It is almost entirely limestone, with an abundance of distinctive karst features such as a huge polje, many smaller dolines or swallow-holes, and bare limestone pavement and cliffs; there is hardly any permanent surface water.

At an altitude of about 600 m on the southern slope, Delphi is a good starting point. Once, the ancient site was exceptionally flowery; now it is less so, but it's still good, and the terraces and cliffs outside the site are usually spectacular in April and May. Densely flowery pink bushes of Judas Tree tower over clumps of golden alkanets, several species of Golden Drop, beautiful clumps of silvery-leaved mulleins, spreading patches of endemic bellflowers, *Campanula topaliana* ssp. *delphica*, all mixed in with Red Valerian, blue-purple vetches, the endemic Swainson's Woundwort and drifts of yellow or blue fenugreeks. In autumn, masses of pinkish-purple Colchicums, late-flowering crocuses and glorious yellow Winter Daffodils can be seen.

Higher up, almost any open area between 1,000 and 1,200 m is likely to be a riot of colour in May. In recent years the intensity of grazing by wandering flocks of sheep and goats has declined and the floweriness has increased. A typical area may be dominated by golden cushions of Spiny Broom with an equally spiny

white milk-vetch, big patches of white Snow-in-Summer, carpets of red or white Southern Bartsia, sprawling clumps of intense blue Herbaceous Periwinkle and the glorious pink Prostrate Plum. Between the bushes grow several species of grape hyacinth, dwarf irises in blue and yellow, masses of purplish orchids such as Four-spotted Orchid, dense patches of bright scarlet Pheasant's Eye, sheets of Pink Hawksbeard and a hundred other species. Towards the snowline, especially in April, are dazzling displays of three or more species of crocus – a fabulous sight, especially in sunny weather when they all open. Higher still, at around 1,800 m, depending on the time of year and the snow cover, you might see great clumps of pink Bigroot Cranesbill, lovely little gems like Burnt Candytuft, Aubretia, Alpine Squill, Oriental Bugle, Greek Fritillary, more irises and orchids, and beautiful big purplish-pink clumps of Gargano Deadnettle. Beyond the end of the road lies the territory of the rare and special plants of Parnassus, though it's too rocky and windswept up here for anything to occur in spectacular numbers.

Opposite: Dense patches of scarlet Pheasant's Eye can be found in the higher grasslands on Parnassus.

Below: A wonderful sward of crocuses, made up of three species with hybrids, erupts into flower as the snow recedes in spring.

The Mani Peninsula

INFORMATION

Location | The southernmost part of mainland Greece, the central 'finger' of the Peloponnese.

Reasons to go | Wonderful displays of early spring flowers. Many endemics, numerous species of orchids often in abundance, in superb scenery with historic villages.

Timing | Peak displays from mid-March to mid-April in the lowlands, with steady progression of mountain flowers through spring and summer.

Protected status | Official protection is very poor and patchy, and few areas are wholly protected.

Opposite: A spectacular display of red Peacock Anemones and other spring flowers near Kardamili on the Mani peninsula.

A warm sunny day on Mani in early April is like a day in Paradise. Abundant and beautiful flowers of many species are combined with a spectacular landscape overlain by a long history of cultivation and habitation. You feel the hand of history here as much as anywhere in the world, and it is still wonderfully unspoilt. The peninsula is a wild and rocky area, with few forests and a barren appearance for much of the year, especially in late summer. But in March and April, especially after a wet winter, it is transformed into a beautiful rock garden.

Roadsides become endless corridors of colour between rocks and stone walls, dominated by the magenta flowers of Calabrian Soapwort, dusty pink cranesbills, scarlet Peacock Anemones, pinky-purple stocks, blue lupins, Pink Hawksbeard, and golden yellow sheets of fenugreeks. Many olive groves here are too rocky to be cultivated, and not worth irrigating or spraying, and they are astonishingly flowery. Orchids are everywhere, including swathes of more common species such as the Italian Orchid or Giant Orchid, with masses of the less common spider orchids such as the striking Reinhold's Orchid, endemic Horseshoe Orchid or Spruner's Orchid, Yellow Bee Orchid and Horned Orchid. There are many other lovely bulbous plants, too, among them three species of tulip, two species of fritillary (including the rare endemic *Fritillaria davisii*), both white and yellow stars-of-Bethlehem, blue spikes of the hyacinth-like *Bellevalia dubia*, several narcissi and much else.

A different range of flowers grows on cliffs: huge clumps of an endemic rock bellflower, growing with the rare endemic woundwort *Stachys canescens*, rosettes of an endemic knapweed, beautiful clumps of Tree Spurge, so reminiscent of Africa, and many other lovely flowers. Shaded rocks are home to the striking blue-flowered endemic gromwell *Lithodora zahnii*, an early flowering blue endemic squill, clumps of caper, great curtains of Joint Pine (at its best in autumn when the red berries ripen), a pretty, large-flowered white meadow-rue and many others.

Although the peak of flowering in the lowlands and up to about 1,000 m is from late March to late April, the season continues higher up the slopes until July, and the higher parts of Taigetos are best visited from late June onwards. Mani has flowers in every month, and late autumn and winter can be very rewarding, as first the cyclamens flower, followed by a succession of crocuses, colchicums, golden-yellow *Sternbergia* species, winter-flowering narcissi, rare snowdrops, and eventually the early spring orchids and bulbs. The area is dotted with lovely old villages and tiny byzantine churches; Judas Trees and fruit blossom light up the hillsides, and mountains are always in the distance, often snow-capped. It's a fabulous place.

Lesvos

INFORMATION

Location | The eastern Aegean Sea, just off the coast of Turkey.

Reasons to go | Lovely unspoilt island, with spring coastal flowers, ancient olive groves and mountain flowers.

Timing | At its peak in April and May, though something of interest for most of the summer and autumn.

Protected status | Mostly unprotected; there are three Natura 2000 sites, of which Mount Olympus is the most important for flowers.

Above: Even the beaches of Lesvos are covered with flowers in the spring.

Conjure up a vision of a Greek island in spring, with flowery olive groves, donkeys, little fishing boats, friendly tavernas and an intense blue sea, and it may well be Lesvos (or Lesbos) that you're imagining. It is quintessentially Greek, very beautiful and flowery, and largely unspoilt. Ecologically, though, it is quite Turkish, lying only 10 km from Turkey, whereas the nearest part of the Greek mainland is some 200 km away.

The rich flora of about 1,600 species is particularly east Aegean or even Turkish in its affinities, with just a few endemics. Unusually for Greece, much of the rock is of volcanic origin, with only small patches of limestone. This makes for fascinating countryside, much of it only sparsely cultivated with scattered fields, barns and walls. Another distinctive feature is the expanse of olive groves – said to comprise over 11 million trees. In some parts of Greece, such as Crete, olive groves are disappointing to the flower-seeker, rather barren irrigated cultivated fields with trees. But on Lesvos most of them are old, with lovely pasture beneath the trees and alive with birdsong and butterflies in spring – more like Spanish oak *dehesa*.

Some of the most flowery habitats here are the sandy strandlines, wherever they have not been developed or driven over. At their best, usually in April, they can be a kaleidoscope of colour stretching away towards the shimmering blue Aegean. Flowers to be found here include masses of yellow fenugreeks, pink and purple stocks, masses of the common but pretty red Mediterranean Catchfly (also delightfully known as Pink Pirouette), Purple Bugloss, scarlet and orange poppies, electric-blue Cretan Eryngo, and hummocks of Spiny Knapweed, Greek Spiny Spurge and Narrow-leaved Milk-vetch. Less stable dunes, with more open sand, are often a threadbare carpet of intense colours – the deep scarlet and black of a special poppy *Papaver nigrotinctum*, gentian-blue patches of Dyer's Alkanet, masses of pale lemon-yellow Sea Medick, the lovely Yellow Horned Poppy, mayweeds, bellflowers, broomrapes and scattered spikes of orchids such as Holy Orchid or Pyramidal Orchid. A little later in the year, the striking scented white flowers of Sea Daffodil appear.

The olive groves can be beautiful in spring as Crown Anemones and Peacock Anemones in almost every conceivable shade burst forth to be followed by orchids, Dutchman's Pipe and many other flowers. One of the most reliably flowery parts of the island is the northern slope of Mount Olympus, around and above the town of Aghiassos. Ancient paved stone tracks wind through olive groves with a wealth of orchids and other flowers. In open woods, there are tulips,

Above: The limestone pastures of Mount Olympus have a wonderful spring show of Crown Anemones and other flowers.

red paeonies, pinkish-purple cyclamen, some special fritillaries (*Fritillaria pontica* ssp. *substipelata*), bright yellow celandines, white Stars-of-Bethlehem and many others. Earlier in spring, wild Broad-leaved Snowdrops bloom here, followed by clumps of the extraordinary Comper's Orchid. In autumn, you can find masses of two species of cyclamen, together with autumn crocuses and the beautiful yellow Winter Daffodils. It's a beautifully unspoilt area.

92 The mountains of western Crete

INFORMATION

Location | Two limestone ranges in the western third of Crete.

Reasons to go | Marvellous displays of spring flowers, extending into early summer, in a spectacular landscape dissected by deep gorges. All the mountains of Crete are good for flowers, though these two areas are reliably the best.

Timing | Peak displays from late March to late April in the mid-altitude areas, with steady progression of mountain flowers through spring and summer.

Protected status | Official protection is very poor and patchy, and few areas are wholly protected; the Samaria National Park protects a small part of the White Mountains.

Opposite: The mid-altitude pastures of the Kedros Mountains are spectacularly flowery in spring.

The sun-soaked island of Crete lies like a stepping-stone midway between Europe and Africa, and not much further from Asia. It is an extraordinary place, with a long history of human occupation, moulded by endless wars of possession and persecution, though nowadays it is peacefully Greek, extraordinarily beautiful, and home to an abundance of flowers. It has been isolated from nearby landmasses for several million years, and having escaped the main effects of the last ice ages, its flora has evolved in isolation over this long period of time. Just under 2,000 species of flowers are native here, of which at least 170 are endemic to the Cretan area.

The lowlands display an abundance of flowers from about mid-March onwards, but are more developed and more intensively farmed than some parts of Greece. By contrast, the mountain areas, above about 1,000 m, are more unspoilt, and the combination of snow and rain in winter followed by intense dry heat in summer produces epic displays of flowers in spring. The areas selected here for their sensational combinations of massed displays and abundant rarities are the White Mountains, or Levka Ori, due south of Chania, and the Kedros mountains, or Oros Kedros, due east of Spili, though many other areas can be very flowery too.

The White Mountains are amongst the highest mountains in Crete, with many peaks over 2,000 m remaining snow-covered into the summer. They are bare, unforgiving mountains with difficult access except for one or two points. A good way to appreciate them is to start from Omalos. The village lies in a basin, the Omalos plateau, which is a fine example of a polje. It has a long history of cultivation and pastoralism, with a general decline in flowers in recent years, but in good years – when snow and rain extends into March – the displays of flowers are wonderful, with masses of pink and silvery-white tulips vying with blue and purple Crown Anemones, sand crocuses, yellow stars-of-Bethlehem, Gargano Deadnettle in great pink patches, bushes of silky pink *Daphne sericea* amongst the rocks, and occasional patches of the striking white endemic paeony *Paeonia clusii*. From the end of the road, at the top of the Samaria Gorge, a wonderful but steep path leads south-westwards up Ginghilos, revealing spectacular displays of endemic and rare mountain plants as you toil your way upwards. Masses of endemic crocuses and squills flower early, followed by such gems as the intense blue *Anchusa caespitosa*, white-flowered saxifrages, Golden Drop, several orchids, particularly the cowslip-yellow Few-flowered Orchid, clumps of lilac-blue aubretia, and blue and white Heldreich's Anemones. My favourites, though, are the

stunning cushions of a dwarf plum, *Prunus prostrata*, whose intense pink flowers cover the tiny gnarled branches as they hug the rocks.

Running southwards from the mountains is a series of deep gorges, of which the biggest and best known is the Samaria Gorge. Escaping the drought of summer and the browsing of goats, these gorges are particularly flowery, with many endemics. At times, whole cliffs stretching up into the mountains can be covered in special flowers such as endemic bellflowers, Giant Fennel, Globe-flowered Coronilla, and the extraordinary Dittany (an endemic cliff-dwelling herb, related to marjoram, and eagerly sought as a tea and medicine). On flatter areas patches of wild white paeonies, drifts of endemic cyclamen, and many orchids can be seen. These wonderful specialist cliff-dwelling plants are known collectively as chasmophytes, and Crete is particularly rich in them.

Further east, just east of the attractive town of Spili, lies the Kedros massif, another astonishingly flowery place. Here, fields are coloured scarlet with the endemic tulip *Tulipa doerfleri*, surrounded by hills with an exceptional abundance and diversity of orchids as well as three species of iris, two other tulips, fritillaries, anemones, and many other flowers – an extraordinary place, best visited in the first weeks of April.

Opposite: In a few places near Spili, the arable fields turn red with the flowers of an endemic tulip in April.

Above: A lovely clump of Crown Anemones.

Below: A wonderful display of early spring flowers on the Omalos plateau.

96 The Pontic Alps

BY ANDY BYFIELD

Opposite top: The distinctive china-blue flowers of Rosen's Squill growing in masses on a mountain slope.

Opposite bottom: The strikingly beautiful pink form of Primrose is abundant in shady places throughout the area.

The length of Turkey's Black Sea shoreline is bordered by mountains, but they are at their most majestic close to the Georgian border, where dramatic granite peaks reaching nearly 4,000 metres are clothed in lush forest and flower-rich pastures and meadows. This is Colchis, land of the Golden Fleece, commemorated in the scientific epithet of so many plants in the region, *colchica*. The flora is exceptionally rich: roughly 2,500 species are recorded, of which 80 are Turkish endemics, and 300 are nationally rare.

The granite peaks of the Kaçkar Mountains form the core of the site, but extensive basalt and igneous rocks also outcrop here, particularly at lower altitudes. Proximity to the Black Sea means that the climate at low altitudes is mild, and rainfall is heavy, reaching as much as 2,500 mm to the east of Rize. The botanical season here is long, with dramatic displays of flowers at differing altitudes according to season. Spring comes early near sea level, where forest clearings and hazel groves boast fabulous displays of hellebores, the brilliant blue *Omphalodes cappadocica*, and magenta and pink *Cyclamen coum* in countless thousands. Primroses deserve particular mention, for at low altitude they are represented by the rich cherry pink subspecies *sibthorpii*, giving way to the more familiar yellow subspecies *vulgaris* at higher altitudes. At least five species of snowdrop occur, including the recently discovered *Galanthus koenenianus*.

But for the best floral displays the forest clearings, hay meadows, and rock outcrops offer a greater diversity of plants, at their best from mid-May to mid-June. Clearings and meadows have abundant cranesbills (including the dramatic magenta *Geranium psilostemon*), globe-flowers and a range of beautiful louseworts. Woodland edge and grassland orchids are abundant here – including the strange greenish-purple *Steveniella satyroides*, and a cream-flowered relative of the Globe Orchid, *Traunsteinera sphaerica*.

As trees give way to open moorland, the intervening scrub zone is very rich, dominated by birch, buckthorns and a range of rhododendrons of which the most beautiful is the richly scented, Golden-flowered Azalea *Rhododendron luteum*. Look here for the blue and white columbine *Aquilegia olympica*, the lemon-flowered *Paeonia wittmanniana*, and a range of yellow lilies during June.

The moorland and alpine zone is at its best from mid-June through to the end of July, with extensive stands of the low, milk-white Caucasian Rhododendron, whilst the open alpine pastures are a mass of betony, bellflowers, gentians, mountain pansies, and a good range of primulas, including the imperial Purple

Oxlip, and drifts of a drumhead primrose (*Primula auriculata*) in damp spots. The rich bulbous flora includes white and yellow stars-of-Bethlehem, and blue squills (including the dramatic turquoise blue *Scilla sibirica* ssp. *armena*, and the lovely *Scilla rosenii*). But for many, the most splendid sights at this time of year are the drifts of the squat, reddish-brown Caucasian Snake's Head. Visit in the autumn for crocuses in their thousands – notably the only yellow-flowered autumn crocus *Crocus scharojanii*, and the wispy, elegant white flowers of *Crocus vallicola*. More obvious still are the drifts of the dramatic Meadow Saffron, parent of so many garden hybrids.

With tea plantations, majestic forests, dramatic banks of rhododendrons, drifts of lilies, and flower-filled alpine pastures, the Pontic Alps have more of the feel of the Himalayan mountains than of European ranges further west; they are worth the long journey.

The high grazing pastures of the Taurus Mountains

INFORMATION

Location | Southern Turkey, just inland from the Mediterranean, roughly between Antalya and Adana, extending west into the Bey Mountains and east into the anti-Taurus.

Reasons to go | Lovely high-altitude displays of spring flowers, including many bulbs; beautiful unspoilt scenery; historic villages; and a rich fauna of birds and butterflies.

Timing | Best from late March to mid-May, though there is usually something to see between early March and late October. Varies considerably according to winter snow, spring temperatures and altitude.

Protected status | Protection is very limited and patchy. Some of the best sites lie within the new Gidengelmez daglari National Park, though levels of protection are not yet clearly defined.

Opposite: Massed ranks of Giant Snowdrops flowering just as the snow melts on the high grazing lands.

The Taurus Mountains extend for almost 600 km across the southernmost part of Turkey, separating the high Anatolian Plateau from the Mediterranean, and, nowadays, separating the bustling tourist scene of the coast from the more traditional way of life of the interior. They are high mountains with many peaks over 3,000 m. This is a wild area, with few towns and a generally low population density, but it is not a wilderness – these mountains have been lived in for millennia, and domestic stock have been grazed in the high pastures since time immemorial.

It is typical of both Turkey and much of Europe that this long pattern of grazing has caused the natural treeline to be lowered. Active clearing of trees, and prevention of regeneration by grazing and browsing, leads to a mosaic of grassland and woodland, with many high pastures to be found well below the natural treeline. Known as *yaylas* or *yaylasi*, these high pastures are usually wonderful places, frequently amongst spectacular mountain scenery. Because they are ancient, there has been ample time for them to be colonized by plants from true alpine pastures, or from woodland clearings, and most of them now boast a lovely mixture of flowers.

As the winter snow recedes, usually in March or April, the pastures begin to be exposed and plants, especially bulbs, begin to flower. Sheets of golden Winter Aconite, usually in its large-flowered narrow-leaved form, *Eranthis cilicicus*, are common, often growing close to masses of yellow crocuses, bluish-purple crocuses, and hybrids between the two. With them, you will often see quantities of confusingly similar spring-flowering colchicums such as Three-leaved Colchicum. Close examination will always separate the two groups, as crocuses have three stamens, whereas colchicums have six. At about the same time more yayla gems begin to flower – the snowdrops. Several species are found here, often in abundance, but the most frequent in spring is usually the lovely Giant or Greater Snowdrop with showy large green and white flowers, peeping out from rock-crannies or even pushing through the snow. Other common species include several blue grape hyacinths, both yellow and white stars-of-Bethlehem, the intense blue Alpine Squill and the pretty pinkish-purple spikes of Wendelbo's Corydalis. Rarer species include maroon and brown fritillaries, dwarf blue irises such as *Iris galatica*, or bright red tulips.

Some places, perhaps where the soil has been slightly disturbed by animals, may have an interesting collection of different plants including the striking tall

yellow barberry-relative, *Leontice*, the intense scarlet of several species of Pheasant's Eye, and the bright yellow of a buttercup-relative, Ceratocephalus. All have succesfully established themselves in the lowlands of east Europe and west Asia as weeds of arable land, but this is probably where they started out. A bonus of these high pastures is the frequent presence of ancient gnarled trees, often close to their altitudinal limit – here, they are most likely to be stately Cedars of Lebanon, Cilician Fir, or enormous old junipers.

There are yaylas throughout these mountains, most commonly at or above 1,100 m, and almost all have an interesting flora. My favourites, with lovely displays of flowers in spring, include Gembos Yayla, above Ibradi, and the extensive Suleymaniye Yaylasi, both fabulously flowery, grading upwards into dramatic limestone karst scenery.

Though the most spectacular flowers are usually in spring, visits at other times are also rewarding; some plants are beginning to flower by February, many flower in May and June, and there are quite a few autumn-flowering bulbs.

Opposite: A wonderful display of wild Grape Hyacinths in an old cemetry near Ibradi, with wild plums in flower beyond.

Below: A spectacular mass of the large-flowered form of Winter Aconite flowering along the snowline at Gembos Yayla.

102 # The Republic of Cyprus (South Cyprus)

Location | The eastern Mediterranean, south of Turkey.

Reasons to go | Many endemic flowers, several species of orchids often in abundance, and good displays of spring flowers.

Timing | Peak displays from early March to mid-April in the lowlands, with steady progression of mountain flowers through spring and summer.

Protected status | Protection is poor and patchy, and few areas are certainly protected, though the Troodos Forest Park is reasonably safe.

Opposite top: Masses of lovely Crown Daisies on the Akamas Peninsula in early spring.

Opposite bottom: Cyprus has a wonderful array of orchids, such as these stately Giant Orchids in the hills above Paphos.

Overleaf: A wonderful field full of Turban Buttercups in March on Akrotiri Peninsula.

Cyprus is split in two politically, with the southern two-thirds Greek-speaking, and the northern part more closely allied with Turkey. Only the southern part is included here, the Republic of Cyprus, not because the north is uninteresting, but the two parts of the island cannot realistically be visited as a unit, and the south has the stronger claims thanks to the presence of the high Troodos mountains, the Akamas peninsula, and the extensive grasslands around the southern salt lakes, none of which has close equivalents in the north.

The number of flowering plant species on the whole island amounts to about 1,800, of which some 140 are endemic, most of these occuring in the south. Cyprus hides its riches well, and you need to search for them, partly because so much has been developed recently for tourism or agriculture, but they are worth seeking out.

The fascinating Akamas peninsula is wild and largely deserted thanks to its harsh topography and geology, and to its recent history as a military area. Its complex geology is dominated by volcanic rock, with areas of serpentine, limestone and other formations. Although proposed as a national park, it has been 'proposed' for many years, which means that the tide of development washes ever closer. Nonetheless, it's a wonderful place, full of plants and secret places.

Most plants flower early here, and late March or early April is the best time to see them. The orchids are marvellous, with quantities of widespread species such as Roman Marsh Orchid, Bee Orchid, Dense-flowered Orchid, Giant Orchid and Syrian Orchid, and an abundance of rarer species and endemics such as the stately *Orchis punctulata*, the Lapethos Ophrys, Bornmueller's Ophrys, and the beautiful Elegant Ophrys. An excellent place to see most species is around the Smygies picnic site, where there are also good populations of rare endemics such as the Akamas Alison, Aphrodite's Knapweed, and the distinctive *Thymus integer*. The north slopes of Akamas, where they tumble down to the sea, have fabulous displays of Persian Cyclamen, Crown Anemones, Turban Buttercups and many other spring flowers jostling for space. The lovely deep red endemic Cypriot Tulip is quite common in patches, while the southern slopes, west of Lara, have masses of the striking magenta near-endemic *Gladiolus triphyllus*.

The Troodos Mountains occupy a huge area in the centre of the island. They are almost entirely made up of hard volcanic and plutonic rocks, including some excellent examples of ophiolites (uplifted sections of the earth's crust), quite different from the limestones of most Mediterranean islands. The high areas have an abundance of special endemic plants, which begin to flower in March, but peak later in the summer. An early trip will reveal such treasures as the wonderfully named Lady Loch's Glory of the Snows glowing blue in the shade of pines, the rare and attractive Kykko Buttercup and Troodos Buttercup, masses of lovely Purple Rock-cress on every cliff, the distinctive pinkish Plutonian Chamomile, great clumps of Aphrodite's Spurge, often pushing up through the snow, groups of the tiny Cyprus Crocus close to the snow line, and clumps of the stately purple Troodos Orchid under the pines.

The Troodos are also home to the endemic Cyprus Cedar (not quite confined to the eponymous Cedar Valley, but very restricted nonetheless), and some wonderful ancient Black Pine and Stinking Juniper forest at the highest levels. Later in the year, there are many more orchids, the Troodos Golden Drop, several endemic Alisons, the striking red Cyprus Broomrape, and many others. Over a third of the endemic plants of Cyprus are found only in the Troodos Mountains.

Finally, it's worth visiting the vast areas of grassland around the salt lakes on Akrotiri or south of Larnaca in late winter or early spring. Many orchids grow here, often in great abundance, including the rare endemic Kotschy's Orchid.

106   Kitulo National Park

BY ROSALIND SALTER

INFORMATION

Location | Southern Highlands of Tanzania near Matamba, 100 km from Mbeya. The main route is via Chimala from where a dirt road leads up towards the plateau. Only accessible by 4WD.

Reasons to go | Exceptional displays of flowers in a remote and beautiful landscape. Charismatic and range-restricted birds, invertebrates and reptile species.

Timing | Of interest almost at any time between late November and April, but most colourful from December to late February.

Protected status | National park since 2006.

Opposite top: The attractive orchid *Habenaria occlusa* in Kitulo.

Opposite bottom: Wet grassland in February with yellow *Berkheya echinacea*, vermilion red orchid and globe thistle framed by Matamba ridge.

This floral paradise is referred to by locals as 'Bustani ya Mungu' or 'Garden of God', and known to botanists as the Serengeti of flowers. Arriving here you find yourself on top of the world, in a sweeping landscape reminiscent of the Scottish highlands. Perched at around 2,600 metres, Kitulo plateau lies between the Kipengere and Livingstone Mountains, where on volcanic soils it supports the most important montane grassland community in Tanzania. It is internationally regarded as an important centre of plant endemism.

The journey to Kitulo is an adventure, passing through remote and stunning scenery. The initial ascent takes you through *Brachystegia* woodland, which in November is bursting with autumnal colour. Between November and April, Kitulo hosts a breathtaking display of flowers, most notably orchids, many of which are rare and endangered. The jewel of Kitulo is the Numbe Valley which comes alive with millions of purple *Aster tansaniensis* in late November, marking the start of the rainy season. Other commonly occurring but distinctive species found here include Edelweiss look-a-like *Alepidia peduncularis*, everlasting *Helichrysum herbaceum*, parrot-beaked *Gladiolus dalenii* and the elegant white *Delphinium leroyi*. In damper areas, species such as the delicate *Geranium incanum* and scabious *Cephalaria pungens* dominate. The striking endemic red-hot poker *Kniphofia paludosa* and *Lobelia mildbraedii* are found in marshy areas, where they are visited by montane Marsh Widowbirds and Malachite Sunbirds. Streams run along exposed basaltic seams, where unusual and attractive species found on the fringes include a sundew *Drosera madagascariensis*, the dainty yellow bulb *Xyris obscura*, and the orchid *Cynorkis anacamptoides*.

From January, an outstanding diversity of orchids appears, including the pretty *Disa welwitschii*, which litters the Matamba ridge. This spectacular rocky ridge forms the backbone of the park and supports numerous species of balsams, proteas, aloes and heathers, as well as the exquisite and rare iris *Moraea callista*.

From Matamba ridge to the opposite side of Numbe valley you are in the main plateau area where, with luck, you may get a rare glimpse of Denham's Bustard stalking among many beautiful blooms including canary-yellow *Moraea tanzanica* and the nodding heads of *Clematoptis uhehensis*.

From the edge of the plateau you get wonderful views over the Livingstone Mountains, whose forests, dripping with lichens and epiphytic orchids, are home to the elusive Kipunji – Africa's first new genus of primate in 83 years discovered in 2003. The discovery of this primate, endemic to southern Tanzania, highlights the importance of the Southern Highlands as an area of high biodiversity and endemism, which is threatened by a rapidly increasing population, illegal logging and an illegal trade in orchid tubers, known as *chikanda*. In 2001, the Southern Highlands Conservation Society, documented for the first time the harvesting and trade of orchid tubers in Tanzania, involving 45 species, many of which are nationally and locally endemic. It is hoped that the intensive harvesting of the orchid tubers in Kitulo will cease and in turn the local economy will be boosted by the promotion of floral and wildlife tourism.

Still relatively unknown, Kitulo is one of the few remaining places in the world where you get a true feeling of remoteness and wonder. A flower lover could never tire of this place but you will need to take a 4WD, your walking boots and an umbrella!

108   The Namaqua Desert: Goegap and the north

I knew about Namaqualand for decades before I got to visit it, and it was always spoken of in hushed and reverential tones – the holy grail of world botanical hotspots. Now, I've been three times, and can confirm the rumours: at its best, it's fabulous. The Namaqua Desert is so huge that it's divided into two entries here to simplify descriptions, corresponding roughly to the northern part and the southern part. It is a winter rainfall desert, with the low annual rainfall, rarely exceeding 150 mm, coming mainly between March and August. In some years – often for several years at a time – there may be hardly any rainfall.

Botanically, the Namaqua Desert is exceptional, with over 3,000 species of plant native here, far more than any comparable desert elsewhere, and with an unusually high level of endemism – about half the native species are found nowhere else in the world. Amongst this wonderful panoply of flowers are several features of particular interest: the incredibly high number of succulents, numbering about 1,000 species (one tenth of the world's succulent flora), almost 500 species of bulbous plants, nearly all of which have lovely flowers, and those extraordinary little stone-like plants adding a fascinatingly different dimension to the flora. There are probably two main reasons why the Namaqua Desert is so rich in species – it's a very old and stable desert that has not been subject to glaciation or other major climatic changes for a very long time, and its desert climate is relatively benign, with few extreme years.

Despite this extraordinary botanical richness, most people come to see the incredible displays of relatively few annual species, though of course any visit here is immeasurably enhanced by the presence of so many other flowers.

Probably the best single site for flower displays in the region is the Goegap reserve just east of Springbok. There is an excellent visitor centre, café, and botanical garden as a perfect starting point, from where a series of walks or drives take you into the best areas. The craggy desert scenery is spectacularly beautiful, dotted with Quiver Trees, Halfmens Trees, and desert spurges, perhaps with the occasional grazing Gemsbok, or passing Padloper Tortoise to be seen. In August and September, the displays of flowers are astonishing, dominated by orange and yellow gazanias, pale yellow Desert Primroses, pink, yellow and purple vygies, many different orange and yellow daisies, pink or magenta pelargoniums, and masses of bulbs from *Lapeirousia*, *Lachenalia*, *Chlorophytum*, *Moraea* and many other genera. This wonderful sight is made even better by the presence of some lovely birds and mammals as well as reptiles such as the striking Rock Agama. The relatively small reserve alone has over 600 flowers, 45 mammals and about 100 bird species.

Opposite: An Oryx or Gemsbok resting amongst a sea of wildflowers in Goegap reserve.

Above: Spectacular rivers of orange-flowered Glossy-eyed Parachute Daisies in the desert at Nababeep.

Right up against the Namibian border, the Richtersveld National Park offers a quite different experience, with difficult access, few facilities, and extreme remoteness. Rainfall is usually less than 50 mm per year, yet the botanical diversity is amazing, with over 350 endemic species. West from Springbok are some wonderful, albeit unprotected, flower displays. The little mining town of Nababeep turns completely orange in spring with millions of the wonderfully named Glossy-eyed Parachute Daisy, with many other special flowers amongst them, and the road west to Kleinzee on the coast passes through superb heathy hills alive with orchids and other bulbs.

Displays vary wildly in timing and quantity, so use all available sources of information both before arriving and while in the area.

Opposite: A large succulent spurge *Euphorbia dregeana*, with orange daisies and other spring flowers in Goegap reserve.

Below: Striking examples of Quiver Tree amongst masses of spring flowers in the Namaqua Desert near Springbok.

The Namaqua Desert:
Namaqua National Park and the south

Location | South of Springbok, almost as far as Vanrhynsdorp; the Namaqua National Park and Skilpad lies just west of Kamieskroon on the N7.

Reasons to go | Exceptionally colourful displays of early spring flowers, dominated by bulbs and annuals. Quartz fields with 'living stones'.

Timing | Peaks July to September, varying according to the winter rainfall; some plants may put on spectacular displays in autumn (February to April), and a few plants flower best in winter.

Protected status | Rather patchy protection, with one National Park, leaving vast areas unprotected and subject to ranching, mining, and development.

Namaqualand is a vast and fascinating area, not precisely defined, but extending from the Namibian border southwards as far as the Olifants River which reaches the ocean near Strandfontein, in a roughly 100 km wide strip. The northern section is described in the previous entry; this southern section contains the Namaqua National Park, the area known as the Knersvlakte with its fascinating quartz fields, and parts of the Kamiesberg Mountains to the east of Kamieskroon.

The rainfall is slightly higher here than further north, averaging about 280 mm at Kamieskroon and the Namaqua Park, though it varies from as low as 110 mm in some years to has high as 540 mm. The flowers are broadly similar, though many endemics are confined to north or south.

A good starting point for southern Namaqualand is the 60,000 ha Namaqua National Park, lying just west of Kamieskroon. An easily accessible 1000 ha wildflower reserve called Skilpad is located around the park information centre and café. While other areas of Namaqualand vary widely in the number of flowers each year, Skilpad is pretty reliable, and in good years it is spectacular. The reserve consists of a combination of old fields, noted particularly for their displays of annuals, and open flowery unploughed hillsides. In August, the fields turn totally orange with daisies, particularly the Namaqualand Daisy, the Glossy-eyed Parachute Daisy and the Namaqua Parachute Daisy, perhaps with patches of blue felicias or ragworts. Around the café and information centre, the ground may be completely covered with pale yellow Desert Primroses, masses of daisies, and a wonderful selection of other brightly coloured composites including several types of *Gazania* and *Arctotis*, hard to identify to species level, but easy to enjoy. The unploughed areas can be almost as thick with flowers, but are usually more varied, with a delightful mixture of plants such as the striking magenta iris-relative *Lapeirousia silenioides*, sand crocuses, geraniums, spikes of bright yellow cat-tails, and yellow or red vygies amongst the orange and yellow daisies. Here and there, more subtle geophytes can be found when you start to look harder.

One of the most fascinating flowers here, found in all habitats, is the lovely orange Beetle Daisy; at first sight a typical daisy but close inspection may reveal the presence of one or more beetle-like blobs on some of the ray petals. It is thought that these attract genuine beetles which believe that the flower must be a good source of pollen. These desert flowers have to be particularly good at attracting pollinators, as the variable rainfall means that insect populations may lag behind flower populations and be quite rare in years when flowers are abundant. The spectacular colours of most of the flowers have probably evolved to attract whatever insects they can. The remainder of the national park is full of spectacular scenery and interesting desert life, but is generally not so flowery as Skilpad.

Southwards, particularly between Bitterfontein and Vanrhynsdorp, extensive quartz fields gleam bright white in the sun, seemingly barren. They are well worth stopping for as this is the home of the strange 'living stones', barely visible amongst the quartz until they produce their pink, white or yellow flowers in winter or spring.

East of Kamieskroon the Kamiesberg mountains rise sharply above the plain. The rainfall is higher and more reliable here, and there are almost always fine displays of spring flowers towards Leliefontein.

Below: A wonderful selection of *Arctotis*, *Ursinia* and *Felicia* species in the Kamiesbereg Mountains.

Overleaf: An extraordinary display of desert flowers after a rainy winter, looking across to the Gif mountains near Vanrhynsdorp.

116   Nieuwoudtville and the Bokkerveld

116

INFORMATION

Location | Around the small town of Nieuwoudtville, about 50 km north-east of Vanrhynsdorp.

Reasons to go | Astonishing displays of spring flowers, especially bulbous plants in spectacular multi-coloured swathes. Interesting birds and some indigenous culture.

Timing | The peak spring flower displays are from mid-August to October, usually peaking in late August and early September. Autumn (February to April) may also be good.

Protected status | Most of the land is farmed (in some cases in a flower-friendly way), though there are several quite large protected areas.

Right: An exceptional display of spring flowers, especially the yellow spikes of *Bulbinella latifolia*, on renosterveld near Nieuwoudtville.

Opposite: Intensely flowery old fields at Matjesfontein Farm nature reserve, Nieuwoudtville.

As you enter the little town of Nieuwoudtville, on the edge of the Bokkeveld Plateau, a notice modestly proclaims: 'Nieuwoudtville, bulb capital of the world. Visit us.' The notice sits on the edge of a rather bleak flat cultivated field and, even in spring, there seems little to substantiate the claim. But this small settlement sits at the heart of an area of astonishing country, rich both in bulbs and other flowers.

Although close to the southern part of Namaqualand, the flora of this area is quite different. It is higher, damper, and with deeper soils that hold water well into the summer. Both the soil and the climate here lend themselves to agriculture, so the region is quite heavily farmed with many of the flatter areas near the scarp under cultivation. The special plants and spectacular displays are mainly to be found in the vegetation known as renosterveld, named after the Rhinoceros Bush or Renosterbos, though much has been ploughed or cleared.

Almost 1,400 species are found on the plaeau, of which at least 80 are endemic just to this area. The flora is remarkable for the number of bulbs or geophytes, almost all with lovely flowers; just around the town of Nieuwoudtville, there are over 300 species of bulbs, many of them endemic. This huge diversity seems to have evolved to cope with dry summers and hungry porcupines.

Because around 80 per cent of the original renosterveld vegetation has gone, it is not so easy to find good flower areas casually. However, the excellent flower

information centre in Nieuwoudtville gives advice on where to go. A good starting point is the Matjesfontein Reserve, an old farm that lies about 13 km south of Nieuwoudtville, now largely given over to conserving the special flora. The displays here take your breath away, and every field is different. In some places, there may be huge patches of the tall yellow spikes of *Bulbinella*, like fox-tail lilies, perhaps with orange, yellow and blue daisies below, as well as many more subtle bulbous plants amongst them. Other areas are a sea of red and pink, dominated by one or two species of enormous red or yellow sand crocuses, pink and white evening-flowering *Hesperanthus*, several species of *Sparaxis*, three or four different iris-like *Babiana* species, and at least three species of *Diascia* (familiar to gardeners). Perhaps nearby will be a sea of vygies in all colours, while another field will be a mass of yellow, white or orange daisy-relatives such as gazanias or felicias. Throughout this kaleidoscope additional little gems wait to be found – rarities, endemics, or beautifully coloured bulbs, such as some of the dwarf gladioli or tiny purple species of *Lapeirousia*; all wonderful.

Below: The damper soils of renosterveld are especially good for bulbous plants, such as these striking red Sand crocuses, *Romulea sabulosa*.

Above left: Gorgeous pink *Hesperantha pauciflora* vies with other bulbs and annuals near Nieuwoudtville.

Above right: A spike of the lovely pink orchid *Satyrium erectum* growing on damp clay.

Other areas of special interest around the town include Bikoes farm, Orlogskloof, and the town wildflower reserve (which has many species, but is rarely truly spectacular). The Hantam Botanical Garden, established in 2007, is more of a reserve than a garden, covering 6,200 ha that was formerly a flower-friendly farm. It's run by the South African National Biodiversity Institute (SANBI), and offers tours and walks in season. The many species of bird to be found in the area include the Sacred Ibis and the beautiful Blue Crane.

Eastwards towards Calvinia, the climate becomes drier and cooler, and though there is more unspoilt habitat, there are fewer flowers. Westwards, towards the scarp edge, the rainfall increases and the land becomes rockier, gradually changing into fynbos (see p. 120) over sandstone, with some wonderful displays of orchids and some more beautiful geophytes. Northwards, especially on north-facing slopes, it becomes warmer and drier, and there are patches of Quiver Tree with great masses of flowers below including Karoo Violet, shrubby yellow-flowered Bushman's Candles, Drumsticks in white or pink-purple, and pretty blue Sunflaxes.

120 The Fynbos of the south-western Cape

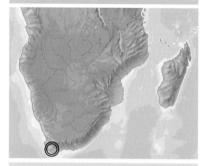
Much the smallest of the world's six floral kingdoms, covering 90,000 km^2, the Cape Floristic Region (CFR) is nevertheless home to over 9,000 species of flowering plant. Most of the flowers can be found in fynbos, dwarf shrubby vegetation roughly equivalent to the maquis and garrigue of the Mediterranean. Although not confined to the CFR, fynbos reaches its peak of development here, and is only rarely found outside the region. It is mainly, though not wholly, associated with bands of sandstone and quartzite on the Cape Fold Mountains.

The displays are not quite as spectacular as further north, but the irresistible combination of exceptional biodiversity and intense floweriness should not be missed. The climate is cooler here, more typically Mediterranean, and the vegetation is more shrubby than in the desert areas, which smooths out the peaks of flowering and reduces opportunities for bare-ground specialists.

The south-western Cape is one of the more populous parts of South Africa, so it is no surprise that fynbos areas are heavily fragmented. Nonetheless, the fynbos is at its best here, and there are some superb protected areas, particularly Table Mountain National Park, the Fernkloof Nature Reserve at Hermanus, and the enormous Kogelberg Nature Reserve, north of Betty's Bay. These three key sites provide a wonderful cross-section of fynbos vegetation with representatives of most of the groups of plants.

A good place to start any exploration of the fynbos is Table Mountain National Park, which brings together most of the high and wild areas of the Cape Peninsula southwards from Cape Town to the Cape of Good Hope. Many exceptional areas of fynbos are near the excellent visitor centre in the south of the park at Buffelsfontein. The general colour and form of the vegetation in spring is established by one or other of the many species of the protea family, such as proteas themselves, conebushes, pagoda bushes including the rare Marsh Pagoda, pinkish spiderheads, or many of the lovely yellowish or reddish pincushions.

Amongst these larger bushes grows a quite extraordinary array of smaller plants, mostly perennials in a dazzling variety of forms: masses of heathers in all colours, the curious black and yellow flowers of Witsenia, yellow Butterfly Lilies, a bewildering number of sorrels in pink, white, red or yellow, fabulous moraeas or Cape tulips in all colours, tall watsonias, many different types of wild gladioli, and intensely coloured babianas. And then, of course, the orchids – dozens of species in many different forms and colours, especially among the spike-like *Satyrium* species and the more spectacular *Disa*, culminating in the spectacular bright red

Right: In August and September, the low shrubby fynbos vegetation of Cape National Park bursts into flower.

Pride of Table Mountain. While in the area, the excellent Kirstenbosch Botanic Garden is always worth a visit.

Further east, the vast reserve of Kogelberg includes fabulous mountain fynbos, occupying much of the peninsula north of Betty's Bay and Kleinmond. This Biosphere Reserve and World Heritage site is home to 1,650 flowering species, of which about 150 are endemic. Rare and spectacular species include Marsh Rose (once thought to be extinct), the strikingly beautiful pink shrubby False Everlasting and the lovely Prince of Wales Heath in white, pink and purple.

Finally, the delightful Fernkloof Reserve, just north of Hermanus, has some nice gardens, a useful information centre, and 60 km of trails. Of at least 1,475 species here, 40 are on the Red Data List, 3 are endemic to the reserve, and 20 or so are endemic to the immediate area. It is possibly the most intensely flowery place in the temperate world.

Opposite: Cape Town's Table Mountain not only has breathtaking views, but also an incredibly rich flora.

Below: The gorgeous orange flowers of a pincushion bush, *Leucospermum cordifolium* in fynbos near Cape Town.

124 The Great Caucasus

Location | Stretching across the whole northern border of Georgia, with substantial parts also in Russia, Azerbaijan, and smaller countries with limited recognition such as South Ossetia.

Reasons to go | Exceptionally rich alpine, montane, and forest flora with many special rare and endemic plants.

Timing | Best from May to July. Flowers begin in March, though access is often difficult then, and continue well into autumn.

Protected status | Official protection is limited, and affected by the political instability of the area.

Opposite: Fields full of Large-sepalled Primulas (the Caucasian form of Cowslip) in the Pasanauri Valley.

Everything about the Great Caucasus is awe-inspiring. It's not simply that the mountains are breathtakingly high, with many peaks reaching beyond 5,000 m, but also that the range is huge, separating cultures and countries, and dominating the landscape of a whole region. Some mountain ranges wear their altitude gently, so that you barely realize you're high up; not the Great Caucasus, where the slopes are unbelievably steep, avalanches and landslides are constant, and vast scree slopes and chaotic boulder fields reach to the sky. You can't help feeling that the whole area is still actively 'under construction'.

Much of the range lies above the treeline, which has been lowered by millennia of grazing, browsing and clearance, and huge areas of open alpine grasslands, cliffs, screes, snowfields and even glaciers provide homes for a spectacular quantity and variety of flowers. The flora of the Caucasus is at least 6,300 species, of which about 1,600 are endemic. Many have been introduced to gardens around the world (and it is certain that many more could be), and most grow in astonishing quantity.

From March onwards the snow melts from the high areas and the first flowers appear, though any new shoots are liable to be completely covered again by snow at any time until well into summer. The hanging green and white lanterns of millions of snowdrops (belonging to six or more species), carpets of the dainty little blue or white Caucasian Anemone, and hillsides of purplish blue pasque flowers (mostly the endemic Violet Pasque Flower) start the show. Soon afterwards drifts of pink, purple and yellow primulas start to flower. Damper slopes, perhaps with less grazing pressure, may have masses of yellow, purple and chequered fritillaries, a sward of china-blue squills and grape hyacinths, or scattered bright golden flowers of the yellow stars-of-Bethlehem, from any one of half a dozen similar species. At the same time of year, cliffs that are free of snow, especially if they are shaded, light up with yellow, white or purple saxifrages (including a number of exceptionally beautiful endemics such as the yellow *Saxifraga ruprechtiana*), intensely flowery cushions of the lovely yellow endemic *Draba bryoides* (far more beautiful than most species of this genus), and the earliest yellow or pink cinquefoils. May brings the first gentians, notably the startling blue clumps of *Gentiana angulosa* heralding a continuous show of flowers that will last right through the summer in all but the driest of years. More gentians, cinquefoils, globe flowers, knapweeds (including some glorious endemics), scabiouses, bellflowers, some gorgeous lilies, masses of Lobed False

Helleborine, monkshoods, marsh orchids and many more in an ever-changing series of displays.

Lower parts of the range, though still high, have a more serene air, dominated by forests of Oriental Beech, firs, spruces and pines, with scattered villages and fields. Though not so spectacular, these are still wonderfully flowery places on a grand scale.

The Great Caucasus range extends through many countries (the political instability of this area is such that it would be unwise to give an exact number), of which Georgia currently gives the easiest access to the mountains. Wonderful high areas are within easy reach of the capital Tbilisi, especially in the Kazbegi and Gudauri regions, and many more can be reached with a little greater effort.

Below: Yellow Stars of Bethlehem with the 14th century Gergeti Trinity Church beyond, high in the mountains above Stepantsminda.

Opposite: A lovely mixture of endemic Georgian Butterbur with Common Coltsfoot at the snowline in the Gudani valley.

The Zagros Mountains

BY IAN GREEN

INFORMATION

Location | From Bandar Abbas on the Persian Gulf to north-western Iran. The regions between Shiraz and Isfahan provide an accessible and excellent area to explore on a more manageable scale.

Reasons to go | Spring displays are remarkable and varied, especially of choice genera such as tulips, irises and fritillaries, with spectacular displays of dionysias.

Timing | April is best, though March in the south is fine, with good shows in May at higher altitudes; June is wonderful near the snowline.

Protected Status | Within this part of the Zagros there is one large national park (Bamu), with protected areas around Kuh-e Dinar and specific sites such as Dasht-e Laleh which protects millions of Crown Imperials.

Opposite top: Masses of *Tulipa bierbersteiniana* colouring wheatfields with the Kuh-e-Kalur mountains beyond.

Opposite bottom: A fabulous clump of pink *Dionysia bryoides* growing on a limestone cliff.

Each April, millions of Crown Imperials fill the valleys of Dasht-e Laleh with an improbable scene, the huge orange flowers covering the ground for nearly 10 km with only occasional gaps where the red soil shows through. Patches of yellowish *Fritillaria persica* and clumps of rich red *Tulipa systola* provide more colour. Crown Imperials are in fact common through the region all the way from the mountains west of Shiraz to Isfahan.

Shiraz provides the ideal starting point for exploration of the area, with a relaxed atmosphere, plenty of good hotels, and numerous cultural sites, while Isfahan provides the perfect culmination as it makes sense to start any visit to the Zagros in the south and move north. In mid-April the dionysias, Crown Imperials and tulips around Shiraz will already be starting to go over, while just 100 km north-west they will be in full bloom, and it can still seem pretty wintry in the high altitude areas around Chelgerd and Aligoudarz. There are several protected sites for Crown Imperials throughout this region and these may bloom anytime between the first week of April and the first week of May, so you can be sure to find them in magnificent flower somewhere.

As well as various colour forms of *Fritillaria persica* there are other special fritillaries; the tubby yellow and mahogany bells of *Fritillaria reuteri* are perhaps the finest of all and form drifts in wet flushes in two of the Crown Imperial reserves, Dasht-e Laleh, near Chelgerd, and Golestan Kuh, north of Daran. Before the Crown Imperials flower at the latter site (a late flowering reserve) the little yellow-tipped bells of almost-black *Fritillaria zagrica* bloom near snows with yellow and red-eyed *Androsace villosa*. Red tulips throughout this region are *Tulipa systola*, drifts of white *Tulipa biflora* look fabulous, and in the north we start to find yellow forms of *Tulipa montana*. Grape hyacinths are varied and abundant, foxtail lilies make spectacular shows on open steppe, and the fabulous *Iris lycotis*, covered with almost black blooms, is unforgettable. The yellow form of *Anemone biflora* is common, though the rarer scarlet form has flowers twice as large.

Semirom is the highest town in Iran, at 2,800 m, with mountains reaching higher on two sides, notably the high limestone ramparts of Pashmaku where the snow lies deep in the strangely eroded mountain. This is the best area to see the signature plant of the Zagros, the dionysia. Around twenty-five species of dionysia are found between Shiraz and Isfahan including several recent discoveries (this region is still poorly known botanically). Many are yellow, but Semirom's claim to fame is that it has three pink forms. Dionysias are cliff or rock dwellers, but there are many places where these plants are easily accessible, quite often by roadsides, and even on large boulders. Purple-pink *Dionysia mozaffariana* is found only on the mountain that dominates the Isfahan road north of Semirom whilst delicate pale pink *Dionysia iranshahri*, with its improbably tight grey cushions, is found on the next mountain west. All around is the incredibly variable and very beautiful *Dionysia bryoides*, a more widespread species, and one that is at home on open rocky steppe as well as cliffs. Its flowers can be almost white or deepest pink. Rarities include orange-yellow *Dionysia michauxii*, found only behind the university in Shiraz, and three species that grow together on one cliff 2 km over the high pass on the Aligoudarz to Shulabad road. Local *Dionysia lurorum* and newly described *Dionysia crista-galli* both have yellow flowers, but *Dionysia zschummellii* is a vibrant purple. It looks as though someone has splattered the cliff with purple and yellow. Altogether, this little-known area is fabulous.

Tien Shan Mountains

130

BY IAN GREEN

INFORMATION

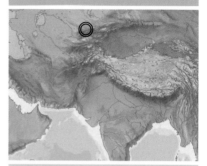

Location | From western China through Kazakhstan and Kirghizia to the borders of Uzbekistan. The most accessible part is between Almaty in Kazakhstan and the Aksu-Dzhabagly Reserve.

Reasons to go | Amazing tulip show in spring with diverse species, colours and forms. A huge range of attractive species in summer including many endemics. Great bird and mammal life, especially in the Aksu-Dzhabagly Reserve.

Timing | The Aksu-Dzhabagly Reserve is botanically exceptional from the start of April through to the start of July, whilst the Karatau area is good from late March until the end of June.

Protected Status | A large swathe of pristine mountain country is protected in the Aksu-Dzhabagly Reserve, and areas to the south of Almaty are also protected. Big Almaty Lake National Park lies to the west.

Opposite: A beautiful mountain meadow with foxtail lilies *Eremurus regelii*.

This is the home of the tulip, and probably the evolutionary epicentre of this fine genus. Around 25 species of variously coloured tulips adorn these steep and beautiful mountains and the surrounding steppes and semi-deserts, the bulk flowering in April. But they are only the tip of the botanical iceberg, for the floral diversity is exceptional and a visit in mid-summer will stretch both your camera and your notebook. Saxifrages and the fabulous *Paraquilegia anemonoides* populate cliffs, whilst high ridges host a variety of stunning alpines. Snowmelt communities last into July, so you can see many of the spring bulbs even during June. The variety of colourful alliums is perhaps unparalleled and there are irises, milk-vetches, *Eremurus* and *Eremostachys* at every turn. Many of these, particularly in the western Tien Shan are found nowhere else, and the nearby Karatau Mountains have one of the highest levels of endemism found anywhere in the world.

Despite its size, the Tien Shan has relatively few places with easy access. Two areas stand out: the former capital, Almaty, and a small village at the western end of the mountains, Dzhabagly, reached either by road or overnight train from Almaty, which has the additional advantage of being only just over an hour's drive from the much older Karatau Mountains, with incredible spring displays of Greig's Tulips, in all colours, as well as Waterlily Tulip, and the lovely pink *Fritillaria stenanthera*. Various roses and endemic fruit trees make a colourful backdrop for bright yellow *Corydalis severzowii* and the rare *Corydalis schanginii* var. *ainii*. The famous Karatau Onion can be found here in its wild state along with various bulbs, and yet more tulips.

When the higher areas above Dazhabagly are still snowbound, white stars of *Crocus alatavicus* and the stunning little iris *Iridiodictyum kolpakowskianum* appear at the lower snow patches. Tulips such as *Tulipa tarda* and *T. bifloriformis* grow with the strange tall fritillary *Korolkovia severzowii*. A little later, the yellow *Colchicum luteum* appears with large white-flowered *Eremurus olgae*. Iris *tianshanicus* and both *Eremurus lactiflorus* and *E. regellii* herald the start of summer. Now the subalpine meadows are lush and Wild Boar, Argali and Brown Bears are feeding on the rich variety. Tall pale blue *Codonopsis clematidea* has a most striking orange and black pattern inside the bells, and louseworts, borages, honeysuckles and milk-vetches abound. Near snow are the two junos, *Iris subdecolorata* and *Iris orchioides*. High ridges are home to rock jasmines, macrotomias and various bellflower relatives, and cliffs have fine paraquilegias.

From Almaty you can take day trips out to Kapchagai Lake or the Kordoi Pass where tulips mix in a fabulous array of strong colours. The Ketmen Mountains are also superb for tulips in spring. Cimbulak, the ski resort above Medeo, is full of wonderful plants in June. You can find accommodation (at almost 3,000 m) at the old Observatory at Big Almaty Lake a little further to the west. A rough track allows sturdier vehicles to continue beyond 3,500 m at the Cosmos station, where the rocky peak above the track hosts many wonderful saxifrages and high alpine specialities.

Opposite: A fabulous clump of the alpine *Paraquilegia anemonoides* growing high in the Tien Shan.

Below: A spectacular display of the globeflower *Trollius altaicus*.

134 The Grasslands of the Tibetan Marches BY CHRIS GREY-WILSON

INFORMATION

Location | About 80 km north-west of Songpan, west of the World Heritage Site of Juizhaigou on the Sichuan-Gansu border.

Reasons to go | An enormous area of breathtaking views in a region steeped in history with an extensive and colourful flora.

Timing | June–July but also September–October.

Protected status | None at present. However, collecting is forbidden in China and visitors should be aware that although they can travel freely now in many parts of Sichuan and other parts of China, and are free to take photographs, they must adhere to the regional laws.

Where the vastness of the Tibetan Plateau abuts the mountains of western China in Sichuan and the neighbouring province of Qinghai there is a large hinterland, the Tibetan Marches, a little-travelled and fascinating area. From the regional capital of Songpan, the road leads north-west towards Aba and beyond over a series of high passes, all exceeding 4,000 m. The region is barely affected by the summer monsoon, but has higher winter precipitation, often as snow.

This regime is reflected in an exceptional flora, with forest restricted to the gullies and wetter slopes, extensive grassland dotted with grazing yaks, and expansive areas of marsh. Although there are scattered small towns and villages much of the region is inhabited in the summer months by pastoral nomadic Tibetans, their small encampments a feature of the area. There are also some very fine Tibetan monasteries, especially those at Aba and Hanguan.

The grasslands, mostly unspoilt by the use of herbicides, stretch westwards across Tibet for a considerable distance, until the combination of altitude and lack of rain cause them to give way to rocky steppe and semi-desert. Because of the lower rainfall in this part of north-western Sichuan, the rich forests and shrubberies seen further south in the province and in Yunnan are absent: rhododendrons, for instance, are never present in such profusion or variety. Instead, more drought-tolerant conifer and birch woodland is found in the valleys,

Right: A gorgeous flowery meadow with primulas at Napahai in the Tibetan Marches.

especially those that run down from the edge of the Tibetan Plateau. On drier slopes, bushes of white spiraea, pink and yellow honeysuckles and several shrubby cinquefoils are speckled with white, cream and yellow flowers. Clumps of white or pink peony (*Paeonia veitchii*) are common and graze resistant. The meadows are a medley of colour in the early summer, a patchwork of white *Stellera* whose short stems are used by Tibetans for paper-making, masses of yellow, white and pink louseworts (in excess of 150 species in western China alone), several species of golden buttercup, blue daisies (*Aster farreri*) and a purple pea (*Hedysarum*) amongst the grasses. On marshy land, amongst various grasses, sedges and rushes, stately daisies (*Cremanthodium bruneo-pilosa*) nod their heads and wave flag-like in the slightest breeze. In the autumn these same meadows are carpeted with blue trumpet gentians, particularly *Gentiana farreri*.

This huge area requires an extended visit, for the distances are great and travel often impeded by delay and roadworks. Yet it is one of the world's great upland grasslands. To the north the area merges into Qinghai province, once included in north-eastern Tibet and culturally similar. Where the road to Jigzhi crosses the border between the two provinces various species of *Meconopsis* poppy catch the eye, large-flowered yellow *M. integrifolia*, upright spiky blue *M. racemosa*, nodding opalescent blue *M. quintuplinervia*, and stunning scarlet *M. punicea*, with satiny flared, skirt-like flowers.

Above: Vast open marshy areas are dominated by stately yellow daisies *Cremanthodium bruneo-pilosa* and louseworts.

136 Zhongdian Plateau

BY CHRIS GREY-WILSON

INFORMATION

Location | North-west Yunnan. Easily reached from Zhongdian (now renamed Shangri-La, but formerly known as Chungtien).

Reasons to go | Beautiful upland plateau with a rich and colourful flora and a wide variety of attractive habitats.

Timing | Late May (especially for rhododendrons) to mid-July, but also September–October.

Protected status | A large reserve at the northern end of the plateau around Napahai is protected.

Although development has taken place around Zhongdian in recent years, much of the plateau remains as it was many years ago and some of the side valleys and surrounding mountains are certainly under-explored. It is a very rich region floristically; indeed this western region of China following the high, steep divides of the Salween, Mekong and Yangtse rivers, the neighbouring province of Sichuan to the north and the eastern and south-eastern region of Tibet (Xizang) has probably the richest temperate flora in the world. Many plants from here are familiar to gardeners (including numerous species of clematis, gentians, lilies, primulas, rhododendrons and roses) reflecting the fact that this was a prime target of early plant hunters. This large plateau, hemmed in by high mountains, lies at an average altitude of about 3,800 m and stretches for some 120 km north to south.

The first impression of the plateau is one of pastoral bliss: small Tibetan hamlets interspersed with wide stretches of meadow grazed by yak, horses, cattle and other herbivores. Occasional hills rise to interrupt the plateau while beyond, higher mountains are thickly shrouded in rich deciduous and coniferous forests. This area, although relatively dry in spring and winter, is subjected to the summer monsoon: rainfall can be very heavy at times between June and September but there are often long spells of dry sunny weather, especially between June and mid-July. The extensive marshy areas of the plateau are enlivened in May by expansive drifts of purple and yellow primulas forming a haze of colour in places. The drier meadows harbour grazing-resistant plants in quantity: maroon-purple flowers of Black Pea contrast with blue *Iris bulleyana*, purple *Incarvillea zhongdianensis*, apple-green *Euphorbia nematocypha* and the yellow ping-pong heads of *Stellera chamaejasme*, all forming a tapesty of interest and colour.

At Napahai, at the northern end of the plateau, the road rises up through hills splashed with white, pink and yellow rhododendrons in the spring while the glades are stuffed full of blue and pink anemones, pink drumsticks of *Androsace spinulifera*, yellow *Trollius wardii* and blue *Aster souliei* merging with primulas in the damper spots. Wild apples are thick with blossom in May, as if from a sudden snowstorm. Amongst the trees and bushes, slipper orchids are found in abundance, and clumps of *Podophyllum hexandrum* bear pink or white peony-like flowers above unfurling mottled leaves.

Roadside screes harbour a different association of plants on dry sun-drenched slopes. The stately blue or purple spikes of *Meconopsis prattii*, often reaching a metre tall, stand out, as do the scarlet heads of Bulley's Rock Jasmine. Many fine walks can be had in this area. Above Napahai the monastery of Jietang Songlinsi is a must-see attraction, much restored in recent years after the deplorable despoliation it received during the Cultural Revolution.

An autumn visit is also well worthwhile. By mid-September the monsoon rains should have eased off and a number of gentians, (particularly the fabled *Gentiana sino-ornata*) in various shades of blue and purple, various monkshoods, delphiniums and a plethora of balsams in pink, purple, yellow and white enliven the meadows.

Opposite: The curious but distinctive flowers of a lousewort *Pedicularis cranolopha*.

Below: A mountain meadow filled with the lovely drooping spikes of *Primula secundiflora*

138 The Kwongan heaths, Western Australia

Location | North of Perth in an area running roughly from Cervantes to Kalbarri on the coast and inland as far as Mullewa and Perenjori.

Reasons to go | Spectacular displays of annuals in wet years, and steady progression of beautiful and unusual perennials in all but drought years.

Timing | Peak displays are normally in August and September, though variable according to latitude and year.

Protected status | Patchy protection in a series of national parks and reserves, though large areas have been cleared for agriculture.

Opposite: A wonderful kwongan display with pink Rough Honey-Myrtle, purple Crinkle-leaved Firebush and yellow *Glischrocaryon flavescens* in spring.

Heading northwards from Perth the vast plains of Western Australia soon open up before you. Once, most of this was heathland, scrub or woodland, but an ambitious programme of clearance for agriculture through the 19th and 20th centuries has reduced and fragmented the area of natural vegetation. Fortunately, many national parks and reserves have been declared, and wide road verges are left uncultivated. These are all now oases of an often spectacular flora.

The flora of Western Australia as a whole is quite exceptional, with around 12,000 species, of which at least three quarters are endemic. The area being considered here has many thousands of native species. Most of these will be unfamiliar to a visitor, and indeed some of the families are unique to Western Australia. In addition, the Mediterranean to semi-desert climate creates a relatively small window of opportunity for flowering, and the displays in spring can be spectacular. In drier more open habitats, or where there has been some clearance, the land can be a sea of annuals or short-lived perennials after a wet winter. In less-disturbed places, such as national parks, low shrubs and small trees dominate, with a massed colourful flowering that lasts for rather longer.

In a typical piece of Kwongan heath, such as in the Alexander Morrison and Kalbarri National Parks, or the Coomaloo Nature Reserve near Jurien, the massed flowering of shrubs in spring is wonderful. Two of the best-represented families are the protea or banksia family (Proteaceae) and the eucalyptus family (Myrtaceae). Low eremaea bushes and honey-myrtles will be covered with masses of beautiful pincushion-like pink, yellow or orange flowers, while lower-growing yellow or pink feather flowers appear in the gaps. My favourites among the Myrtaceae are the extraordinary starflowers, low bushes covered with tightly packed starry flowers in pink, red, purple, yellow, or combinations of these colours. The Proteaceae are represented by many species of banksia with their curious barrel-shaped clusters of yellow, orange or red flowers, many grevilleas in all colours, the rather similar hakeas and a collection of gorgeous coneflowers in pinks, reds and cream. Dryandras are like small banksias but with more bracts around smaller inflorescences, and they are now usually classified with *Banksia*, though very different in appearance. Between the shrubs may be several species of orange or pink sundews, yellowish cottonheads, yellow guinea flowers, and bright blue dampieras and fanflowers.

More disturbed areas – along roadsides and around towns – and areas of lower rainfall with more open vegetation can have spectacular displays of spring

annuals after rains. Seemingly endless carpets of yellow, white and pink everlastings, purple Four o'clocks or Common Parakeelya, Golden Waitsia, Pompom Heads and Pink Velleias amongst others burst forth for a brief flowering before the intense dry heat of summer sets in. The extraordinary Wreath Plant, so-called because each circular cushion produces a ring of bright pink flowers around its circumference seems to be almost confined to bare roadsides.

In the peak flowering season, almost every village and small town here sets up a wildflower information centre to guide you to the best sites. Mullewa, Mingenew, Enneaba, Perenjori, Jurien and several others all vie for your custom, making a visit an easy and pleasant experience. The flower displays are scattered and variable, so it's definitely worth taking up-to-date advice to help in tracking them down – the effort will be well-rewarded.

Above: Purple Starflower in flower in spring near Geraldton.

Right: Masses of Pink Velleia, and other spring everlasting flowers in semi-desert scrub near Paynes Find.

Opposite: Hummocks of Pincushion *Borya constricta*, amongst masses of Pink Velleia, near Yalgoo.

142 The Stirling Range, Western Australia

Opposite top left: The intriguing Common Donkey Orchid *Diuris corymbosa*.

Opposite top right: Cork Bark Honey-myrtle and other flowers on heathland in the Stirling Ranges.

Opposite bottom: Strikingly flowery Jarrah forest with orange *Gastrolobium*, white Mountain-heath and other spring flowers.

In the flat lands of south-west Australia, the high peaks of the Stirling Ranges stand out like a beacon from every direction. Ecologically, they stand out too: satellite photographs show how all the surrounding land has been cleared and cultivated, leaving the national park as an island of beauty and biodiversity in a sea of monoculture. South-west Australia is one of five parts of the world with a Mediterranean climate, where mild wet winters contrast with hot dry summers. The Stirlings follow this general pattern differing only in that they trap more clouds and rain than the surroundings, and the higher parts are cooler. In fact, they are not particularly high, with most peaks well below 1,000 m.

This is a very special area, with over 1,500 native plant species recorded, of which about 85 are endemic. Some of the special plants include the lovely pink Stirling Range Coneflower, Giant Andersonia, Stirling Range Bottlebrush, the creamy-yellow Mountain Kunzea, 125 species of orchid (including the fabled Queen of Sheba Orchid), and 10 species of the strikingly beautiful mountain bells. These last, in the genus *Darwinia* are restricted to Western Australia, and several of them occur on just one or two mountains in the national park. They are named not after Charles Darwin, but after his grandfather Erasmus, an eminent physician and scientist.

Spring comes later here than in the warmer Kwongan heaths (see p. 138). The best displays usually occur in the lowlands from early to mid-September onwards, and they last rather longer here thanks to the damper cooler climate. Under the open Jarrah or Wandoo forest, the flowers can be breathtaking with a sea of orange *Gastrolobium*, white Mountain-heath, yellow Prickly Moses (a dwarf acacia), Needle-leaved Chorizema, several hakeas, and many orchids. In the open flatter areas, vast stretches of low-growing heathy vegetation feature dwarf banksias, Harsh Hakea, Scallop Flower, *Lambertia*, the strange pink stem-hugging flowers of Corky Honey-myrtle, and many orchids and bulbous plants. Higher up, such as on the slopes of Mount Trio, are some wonderful areas of surprisingly flowery dwarfed woodland, with great patches of red or pink mountain bells, the beautiful pink-flowered Stirling Ranges Coneflower, Bell-fruited Mallee with its lovely lemon-yellow flowers and Giant Andersonia, amongst many others.

Roads give good access to these habitats, and in particular it is worth following the Stirling Range Drive from Red Gum Pass Road to Chester Pass Road, taking any opportunity for a detour. Good paths lead up Mount Magog, Mount Hassell, Bluff Knoll and Mount Trio to the higher altitude vegetation.

144   The mountains of South Island

INFORMATION

Location | South of Blenheim and Nelson, around Alexandra, and in fiordland National Park.

Reasons to go | Fascinating high mountain environments with a dwarfed vegetation unlike any other. A palette of muted silver in stunning mountain scenery.

Timing | Best from December through to late January or early February.

Protected status | Mostly unprotected except for sites in the fiordland and Mount Cook National Parks.

And now for something completely different. New Zealand is hardly famous for its spectacular displays of flowers but a few mountain areas in South Island have a fascinating vegetation, unlike anything else, that makes them worth including. These are all alpine sites, above the natural treeline, and each has something special.

The most northerly are the Blackbirch Mountains, rising to a height of 1,700 m above the Awatere Valley south of Blenheim. These cold, bare, jagged mountains have all the characteristics of tundra, and a diverse flora of small alpine plants. But the speciality of the area is that extraordinary group of plants, the vegetable sheep, for which this is the world centre. These members of the composite family, in the genera *Haastia* and *Raoulia*, have become strongly adapted to a difficult alpine climate. They grow into large tight cushions, occasionally up to 8 m across and 1 m high, though usually rather smaller, resembling a silent flock of sheep where they grow in abundance. The largest cushions are undoubtedly many centuries old, as their growth rates are very slow. The combination of huge hummocks of Giant Vegetable Sheep, smaller cushions of Common Vegetable Sheep, and a mass of smaller silvery-leaved white-flowered dwarfed alpines such as White Cushion Daisy and Large-flowered Mat Daisy is irresistibly intriguing and beautiful. I know of nowhere else with a vegetation that looks like this on such a scale. There are other more 'normal' alpines here, too, that give a hint of colour – yellow mountain daisy, blue harebell, Golden Spaniard, and the curious rosettes of Penwiper Plant. This is a fascinating and beautiful environment.

Right: A lovely clump of White Cushion Daisy at 1,400 m in the Black Birch Range.

Further south, in Central Otago Province, there are some even harsher mountains, with a mean annual temperature at freezing point, and fog on 60 per cent of days. The Old Man Range or Obelisk Mountains – just to the south-west of Alexandra – stand out as the most botanically interesting. They are rounded eroded ridges of schists, capped with dramatic tors, and clothed in a carpet of silver and green alpine flowers and lichens. From 1,400 m or so upwards the vegetation is totally dwarfed yet almost completely continuous, with very few gaps or patches of soil. A wonderful silvery mosaic of cushions and hummocks of vegetable sheep, mat daisies, Edelweiss, Rock Cushion, Red Rock Cushion, Alpine Cushion, Psycrophila, and a dozen lichens stretches away into the distance, interrupted only by snow patches and tors.

These are the two most distinctive mountain areas for their displays of plants, but there are many other places of interest, such as the massed ranks of Mount Cook Lilies (actually a large-flowered white buttercup) in valleys throughout the Southern Alps, or the heavily glaciated valleys in the fiordland National Park, including Gertrude Valley, in spectacularly beautiful wilderness scenery.

Top: 'Vegetable sheep' in high altitude tundra in the Black Birch Range, South Island.

Above: Mount Cook Lily and Large Mountain Daisy in the Gertrude valley, Fiordland National Park.

146 Waterton Lakes National Park

INFORMATION

Location | The southernmost part of the Canadian Rockies, adjacent to the US border and Glacier National Park.

Reasons for going | Beautiful montane prairie grassland and spectacular mountains with a rich flora.

Timing | The prairie grasslands are at their best from mid-June through July; higher areas are best from late July through August.

Protected status | National park, with additional areas beyond boundaries managed by agreement. World Heritage Site.

Opposite: Intensely flowery mid-altitude prairie grassland, with Showy Locoweed and Brown-eyed Susan on the edge of Waterton.

The Canadian Rockies are spectacularly beautiful and wild on an almost unimaginable scale. They are also very flowery, especially above the treeline. In general, however, the notably flower-rich places are too scattered and limited in species to feature here. The one area that does stand out is the Waterton National Park, which is exceptional for two reasons: with about 1,000 species it has a much richer flora than areas further north, and the high peaks are complemented by a wonderful apron of beautifully flowery mid-altitude prairies, which extend beyond the park boundaries.

As in most northern American mountain regions, coniferous forest spreads a dark green blanket around the higher areas, reaching up to the natural treeline at 2,000 metres and above. The woods here are more flowery than most, with carpets of striking white Bunchberry, Red and White Baneberry, Thimbleberry, tall elegant spikes of creamy-white Bear Grass, delicate white flowers of Queen's Cup, sprawling masses of Foam Flower, scrambling vines of Blue Clematis dotted with large blue-purple nodding flowers, and orange arnicas. Special delights of the forests here are the gorgeous clumps of Mountain Lady's Slipper Orchids covered with porcelain-white flowers, not to mention the intriguing Sparrow's Egg Lady's Slipper, Striped Coral-root and many other orchids.

Clearings caused by avalanches and landslides support a fine flowery sward of creamy Rough-leaved Alumroot, several beardtongues – especially the lovely bunches of Blue Beardtongue – blue and white larkspurs, Spotted Saxifrage, striking clumps of orange Balsamroot, and astonishing masses of the beautiful Three-Spot Mariposa-Lily growing in tens of thousands.

Around and above the treeline, there are often spectacular displays of a different range of plants. Low-growing Red, White and Yellow Heathers all form huge mats covered with flowers, together with Mountain Avens, Yellow Dryad, and the extraordinary tight cushions of Moss Campion, covered so densely with pretty pink flowers that it's impossible to see the leaves. Damper areas feature beautiful 'white gardens' of several white anemone species, White Globe Flowers, Mountain Marsh Marigold and Western Spring Beauty dotted with patches of golden Snow Buttercups.

The rest of the Canadian Rockies have similar habitats and species, albeit with a decreasing diversity as you head north. Something that sets Waterton apart is the large area of montane prairie grassland, particularly around the northern and eastern parts of the park at altitudes of about 1,200–1,400 m (with the protected

area recently extended well outside the park by agreements with local ranches). At its best, this grassland can be a lovely tapestry of orange Brown-eyed Susan, huge patches of magenta-pink Showy Locoweed, tall creamy-yellow spikes of Mountain Locoweed, blue lupins, masses of Sticky Purple Geranium, Shrubby Cinquefoil covered with golden flowers, occasional orange stars of Wood Lily and a host of other flowers. These grasslands are grazed by great herds of Elk, and are home to many special mammals, birds and insects. They are a vital and beautiful strand in Waterton's wonderful melting-pot of habitats.

Waterton lies adjacent to, and is twinned with Glacier National Park, geologically and ecologically the southernmost extension of the Canadian Rockies. Further north, many areas are worth visiting for their spectacular flowers: some notable places include Sunshine Meadows, Banff National Park (accessible by shuttle bus), Highwood Pass (the highest surfaced road in Canada at 2,220 m), Peter Lougheed Provincial Park, and the high meadows of Mount Edith Cavelle, in Jasper National Park (accessible only on foot).

Opposite: Mountain Lady's Slipper Orchid in the woods near Waterton.

Below: A lovely Hoary Marmot among Western Anemones.

Bottom: Common gaillardia, lupines and other flowers in prairie grassland, Waterton Lakes National Park.

150 Olympic National Park, Washington

Location | The Olympic Peninsula, west of Seattle, Washington.

Reasons to go | A vast quiet area, with wonderful displays of flowers above the treeline, the best temperate rainforest in the USA, and many animals and birds.

Timing | Peak flowers from late June to mid August, usually best in the second half of July, but varying according to amounts of snow remaining.

Protected status | Wholly protected as a national park, mostly also designated wilderness.

A quick glance at the map might give the impression that the Olympic Peninsula is a busy, built-up area, lying so close to the sprawl of Seattle and Tacoma. In reality, this could hardly be further from the truth, as most of the area is remote, high and wild, with few visitors away from the hotspots. The national park extends over 3,734 km² of mountainous country, with just a few roads edging into it, leaving a huge core wilderness area accessible only on foot.

Although quite close to the Cascade Mountains (see p. 152), the Olympic Mountains are quite different geologically, made up of a complex mass scraped from the Pacific Plate as it sinks below the North American Plate that forms the main landmass of North America. This is an area of notably high precipitation – the forests on the western part of the peninsula receive around 3,560 mm of rain per year, while the higher areas average closer to 5,000 mm which falls mainly as snow. The highest point is Mount Olympus at 2,432 m, which has one of the most extensive glacier systems in the USA outside Alaska.

In common with the other mountains of the north-west, you need to climb into the subalpine or alpine zones in order to see the most spectacular displays of flowers. The forests begin to open out at about 1,300 m, becoming steadily more open as you climb further. The high meadows, in the upper reaches of the subalpine forests, can be astonishingly flowery with multicoloured swathes of blue Broadleaf Lupins, White Avalanche Lily, Beargrass, red, magenta and yellow paintbrushes, several penstemons, Red Mountain Heather, gorgeous pink shooting

Right: Big Leaf Maple trees dripping with epiphytes in wet temperate rain forest, Quinault Valley.

stars of several species, a rather striking red form of Coiled-beak Lousewort and the yellow of several arnicas, often against the backdrop of the spire-like Subalpine Firs, and high snowy mountains in the distance.

Higher still, the vegetation becomes thinner, with more bare rock and lichens, but there can still be some fabulous displays. Sheets of creamy Yellow Coralbells, mats of Partridge Foot, fleabanes and asters, Goldenrod, Shrubby Cinquefoil, intense blue penstemons such as Small-flowered Penstemon, mats of the gorgeous magenta-pink Cliff Primrose, lilac Spreading Phlox and Mountain Owl Clover. It is in these higher, more open areas that you are most likely to find the special endemic plants such as the lovely blue-purple Piper's Bellflower, Olympic Onion or Flett's violet.

The easiest way to get to see the flowery parts of the Olympic Peninsula is by driving to Hurricane Ridge at 1,600 m (it's worth visiting the visitor centre here, for current information), and then taking the Old Deer Park Road. There are many opportunities to walk further from these routes. While in the area, it is also worth visiting the stunning temperate rainforests in the western parts of the park, such as at Quinault or Hoh – they are not especially flowery, but they are quite spectacular in other ways.

Top: A spectacular mass of flowers including Broadleaf Arnica, paintbrush and Broadleaf Lupine at high altitude on Hurricane Ridge.

Above: The lovely magenta flowers of Cliff Primrose high up on Hurricane Ridge.

152 Mount Rainier, Washington

Mount Rainier is the jewel in the crown of the beautiful Cascade Mountains, rising to a massive snow-capped, glacier-clad 4,392 m. It is famous throughout the USA and beyond for its wonderful mountain flower displays, and if you only visit one site to see North American mountain flowers, then this should probably be it. Biodiversity is high, with 900 species of higher plants recorded in the national park, but it is the sheer spectacle that ranks highest.

Within the Cascades, the amount of snow often determines the quality of the flower displays, and this is one reason why Rainier scores so highly. The huge snowfall often amounts to 21–24 m, peaking in a record snowfall for North America of over 29 m at Paradise Meadows in 1971–72. It also rains a good deal, though weather in summer is generally good. This huge accumulation of snow, especially on the south and west flanks of the mountain, takes a long time to melt at subalpine levels (and occasionally it fails to melt altogether), which means that the season for the subalpine and alpine plants and their associated animal life is very short. Good summer weather and ample ground water from snow melt and occasional rainfall means the displays are spectacular, as the flowers pack as much growth and flowering as possible into the few months available to them. However, the timing of the best displays can vary, so it is wise to contact the national park before making a long journey. Not surprisingly, the area becomes very busy at peak times, and an early start is advisable.

The two best accessible areas are around Paradise Meadows and Sunrise, both served by excellent roads and visitor centres. Paradise is the most famous site, lying on the south side of the mountain above where the road ends at about 1,650 m. From here, there is an extensive network of trails up into the meadows, treeline forests, tundra and glaciers of this part of the mountain. This is where the snowfall is highest, but the rewards are greatest. The meadows in high summer are a vibrant mass of colour, dominated particularly by Magenta Paintbrush, Avalanche Lily, Glacier Lily, several species of lupin, Broad-leaf Arnica, asters, Jeffrey's Shooting Star, White Mountain and Red Heathers. The high meadows are great places to see Hoary Marmots playing and feeding, often at very close range, as well as the delightful and confiding Golden-mantled Ground Squirrel. If you are lucky, you may catch a glimpse of the strikingly beautiful Cascades Gray Fox, a brown and grey long-coated version of the Red Fox.

At heights of 2,100 m or more, the tundra is often equally spectacular, with some of the above species and also Partridge Foot, Northern and Dwarf

Goldenrod, stunning clumps of the magenta Cliff Penstemon and the lovely bluish Davidson's Penstemon, the vivid orange Cliff Indian Paintbrush, dwarf lupins and Golden Daisy amongst many others. The tundra areas are extraordinarily beautiful, with wide views reaching up to the highest peaks, and the scattered remnants of the lovely Subalpine Fir forest as it thins out at its altitudinal limit. Ptarmigans are underfoot and ravens and eagles overhead – a magical place.

Other parts of the national park are worth seeing, too. The streams or creeks often have a rich flora along the banks, such as the beautiful large pink-flowered Lewis's Monkey-flower and the yellow mats of Large Mountain Monkey-flower. Another fabulous area lies on the eastern edge of the national park, and partly outside it, in the William O. Douglas Wilderness. The whole area around Chinook Pass on route 410 is superb, especially around Tipsoo Lake and Naches Peak. The flora is broadly similar to that of Rainier, though plants such as Cascade Azalea, Tongue-leaf Rainiera, and Columbia Lily are particularly common. The views of Mount Rainier are lovely in clear weather, and there are far fewer people than at Paradise or Sunrise.

Below: A Calliope Hummingbird visits louseworts high on the slopes of Mount Rainier.

Overleaf: Possibly the most flowery place in the world – the astonishing display of flowers at Mazama Ridge, Mount Rainier.

156   The Klamath–Siskiyou area, California and Oregon

Opposite: Spectacular mountain flowers in the Klamath-Siskyou mountains in July.

Where the western parts of California and Oregon meet, a vast and complicated jumble of high mountains extends over a huge area of barely inhabited countryside known as Klamath–Siskiyou. It includes all of the Klamath Mountains, the Siskiyou Mountains and the Marble Mountains, as well as many subsidiary ranges. Geologically, the area is incredibly complex, with volcanic, metamorphic and sedimentary rocks of all types, including large outcrops of serpentine.

This exceptional region is home to over 3,500 plant species drawn from the floras of the Sierra Nevada, Cascades, Coast Ranges and the Great Basin, with about 280 species that are found nowhere else in the world. It is one of best places for flowers in North America, but it also has a special bird fauna with endangered species like Northern Spotted Owl, more wild undammed rivers than anywhere else in the USA, and a whole host of other endangered or endemic animals. It is very special place.

Most of the endemic plants are 'neoendemics', which have separated quite recently from similar species. A few more distinctly different plants are also found here, such as the saxifrage-relative Oregon Bensoniella, or the annual Indian Head-dress. One of the most extraordinary plants of the region is the California Pitcher Plant, or Cobra Lily, which grows in spectacular abundance in some of the bogs, such as on the slopes of Mount Eddy. Though not quite confined to Klamath–Siskiyou, this is undoubtedly where it is at its best; the sight of a whole wet hillside covered with yellowish-green pitchers, perhaps interspersed with white orchids, the yellow spikes of the endemic California Bog Asphodel, or even a few slipper orchids, is quite unforgettable. Other special plants here include several lilies, the tiny yellowish Siskiyou Fritillary, Henderson's Horkelia, Waldo Gentian and Bolander's Onion, to name but a few. It is also one of the four richest temperate coniferous forests in the world, with 30 species of conifers recorded.

Within the region, some areas are especially good for their displays of flowers or exceptional diversity. Mount Eddy, west of Shasta, is the highest mountain in the region, reaching 2,751 m. Its eastern slopes have some wonderfully flowery bogs, flushes and grasslands, noted for their pitcher plants, but home to many other colourful species. The higher parts of Mount Ashland (2,280 m) have a lovely variety of habitats, with superb grasslands and wet areas and a wonderful range of species including endemics such as Mount Ashland Lupine as well as

many more widespread species of monkshood, lilies, orchids, paintbrushes and penstemons, creating a glorious multi-coloured tapestry.

Further north, the 70 ha Kalmiopsis Wilderness is named after a pretty red-flowered endemic shrub, only discovered in 1930. The marvellous flora here includes many rare and endemic species. Castle Lake and crags, west of Shasta, is another famous botanical locality, with lovely high meadows and some special cliff plants. Perhaps the area with the best displays of flowers is the Marble Mountains, west of Yreka, though the most spectacular sites involve quite a bit of walking. It's worth the effort for astonishingly colourful masses of lilies, blue lupins, drifts of Tolmie Star-tulips, orchids, lewisias, half a dozen or more species of beardtongue, bright red columbines and many others in beautiful alpine meadows.

Opposite: A beautiful clump of Siskiyou Mountain Clover *Orthocarpus cuspidatus* on Mount Ashland, Oregon.

Above: Scarlet Gilia, Sulphur Flower, and Owl's Clover on Mount Ashland, Oregon.

160 The Carrizo Plain National Monument, California

INFORMATION

Location | South-west California, about 160 km north of Los Angeles, or 80 km east of San Luis Obispo. A remarkably remote area, with few facilities.

Reasons to go | Breathtaking flower displays on a vast scale, especially after wet winters, including many rare and endemic species; high populations of breeding and wintering birds; herds of pronghorn and many other mammals; and Native American history.

Timing | Late March to early May, though variable – best to check before visiting.

Protected status | Over 1,000 km² is protected as a National Monument, though there are a number of enclaves of private land within and around this.

Opposite: It's hard to believe that the whole of this view in the Temblor Mountains is coloured gold by Hillside Daisies and other flowers.

Come to the Carrizo Plain at any time between June and December and it might seem a rather dull place, dry, brown and featureless. Yet this is one of the most diverse, beautiful and ecologically significant places in the USA, home to an abundance of endangered species and an outrageous display of spring flowers. Its sheer size is remarkable – the plain stretches for 80 km at its longest, and is 15–20 km wide, bounded by the Temblor Range to the north-east and the Caliente Range to the south-west.

Grassland stretches away into the distance, interrupted only by the vast glistening white salt lakes at its heart. The flat plain is almost all former ranch land, once so heavily grazed that it was considered to be useless desert, but now lightly grazed by herds of Pronghorn Antelopes, thousands of Giant Kangaroo Rats, and domestic stock as necessary. In spring, from late March to early May, it bursts into life as millions of flowers of Goldfields, yellow and white Tidy Tips, golden *Coreopsis*, blue spikes of delphiniums and lupins, purple phacelias and many others fill the plain with huge swathes of colour. There is only one good road through the plain, and in April, it feels as though you are driving through endless fields of flowers.

The San Andreas Fault runs right through the plain, and the rocks on either side are of quite different origin, with land west of the fault moving steadily northwards faster than your fingernails can grow. The plain arose when streams were blocked as a result of tectonic movement, and nowadays the relatively low annual rainfall (about 200 mm) all evaporates from the alkaline lakes.

The Temblor Range (from the Spanish word for earthquake), which bounds the plain on the north-east side, has the most intricate drainage erosion pattern imaginable, thanks to its soft rocks and recent uplift. As a result, a huge and varied canvas of bare ground is eagerly filled each spring with annual flowers that bloom and fruit before the slopes dry out in May. After a wet winter, the swathes of colour are visible from many miles away. Whole hillsides are bright yellow with Hillside Daisy, others may be purple with Lacy Phacelia, or pink with the lovely endemic Parry's Mallow. Enormous patches of orange are most likely to be the attractive endemic San Joaquin Blazing Star, though even more vivid orange patches will probably be California Poppies, opening wide in the sunshine. Some slopes or shoulders have vast waving fields of a most unlikely looking plant – the endemic Desert Candle, a member of the cabbage family, but looking more like a tall inflated Tassel Hyacinth, with its curious

Below: One of the extraordinary hillsides on the Temblors, coloured blue and orange by San Joaquin Blazing Star, laceflowers and Hillside Daisy.

Opposite: The floor of the Carrizo Plain can be covered in flowers, such as these Fremont's Phacelia and Goldfields, as far as the eye can see, .

combination of cylindrical swollen stems and short-stalked purple and white flowers.

Almost all of these flowers produce a strong fragrance, and it is a quite magical experience to stand high on a flower-covered ridge, miles from the nearest house, absorbing the sights and smells of this remote and beautiful spot. If you are happy to walk and climb, the possibilities are endless as new hidden valleys are constantly revealed, but it's also possible to see the flowers at close range from highway 58, or from Crocker Springs Road (unsurfaced) which crosses the range en route to the plain from near Taft. The higher, even more remote Caliente Range to the south has different geology, and higher rainfall, producing fine displays of flowers a little later.

The Tehatchapi Range and Antelope Valley, California

INFORMATION

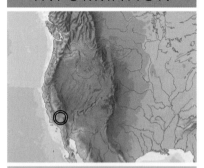

Location | South-western California, roughly midway between Los Angeles and Bakersfield, to the north-east of Interstate 5.

Reasons to go | Exceptionally beautiful and intricate displays of spring flowers over large areas; rich bird life including California Condors and many others.

Timing | Usually mid-March to end of April, but very variable according to weather and altitude.

Protected status | A few parts are protected, but mainly privately owned and currently the subject of detailed negotiations to secure the future of a large part of the area.

Below: Evening Snow, California Poppies and other spring flowers at Gorman in April.

Opposite: A vast mass of blue Bentham's Lupines at Grapevine, at the foot of the Tehatchapi Mountains.

There is an area in southern California where everything seems to come together at one place. The great snowy ridge of the Sierra Nevada, the Coast Range, and the huge flat Central Valley all reach their southernmost points, the Mojave Desert stretches a finger out westwards, the Transverse Ranges cut across eastwards from the coast, and the Sonoran Desert is not far away to the south. Through it all passes the San Andreas Fault, where two continental plates meet. At this nodal point lie the spectacularly flowery Tehatchapi Mountains.

This small range of mountains is roughly between the town of Tehatchapi to the north, and Gorman on Interstate 5 to the south, rising to 2,433 m at Double Mountain, the highest point. Apart from their position at the intersection of so many biogeographical areas, they have also been heavily influenced by the stability of lying largely within the largest block of private land in California. The vast Tejon ranch was originally established by Mexican land grants in the 1840s, and later extended to cover well over 1,000 km². In general, this has led to wise husbandry of the land and conservation of most of its natural features, though the downside is very limited public access, and recently new owners have declared the intention to develop large parts of the estate for housing and industry.

The rich species mix is extravagantly colourful. On the north-west slopes, particularly around The Grapevine on I5, and along the 223 towards Arvin are great swathes of blue lupins, particularly Bentham Lupine, mixed with the pink of Owl's Clover and other species. Where the interstate road crosses the Tejon pass, around the little town of Gorman, the whole south-facing scarp of the range becomes an

intense mosaic of many colours, especially the blue of lupins and phacelias, yellow or orange of tick-seeds and goldfields, intense orange California Poppies, white Evening Snow, pink or purple gilias and owl clovers and many others. In particularly good years, the traffic on the freeway almost comes to a halt as people slow to admire the extraordinary display. It is best seen at close quarters from along Gorman Post Road, immediately east of the town. The newly formed Tejon Ranch Conservancy runs day trips at weekends in season deeper into private land, though places are very limited.

Immediately to the south-east of the Tehatchapis, along highway 138 towards Lancaster, lies the broad flat Antelope Valley. Although much of it lies under agriculture or development, great swathes of wild land dramatically burst into flower in spring, totally dominated by California Poppies. The Antelope Valley California Poppy Reserve protects 730 ha of glorious poppy habitat, but in a good year a sea of poppies stretches in almost every direction around the reserve. Nearby, a fragment of Mojave desert with Joshua Trees is also protected in the Ripley Desert Woodland Reserve.

This is an altogether exceptional area, full of spectacular displays and rich in species – a reminder of what much of California must once have looked like.

Opposite: Intensely flowery grazed pastures in spring on the Tejon Pass at Gorman in the Tehatchapi Mountains.

Below: Breathtaking displays of Californian Poppies at the Antelope Valley California Poppy Reserve.

168   Anza–Borrego State Park and Wilderness, California

INFORMATION

Location | Far south-west California, about 100 miles east north-east of San Diego, extending south almost to the Mexican border.

Reasons to go | Exceptional displays of desert flowers in some years, in fine desert scenery; desert bighorn sheep and good birds.

Timing | Peak flowers from February to April, but quite variable according to rainfall levels. It is essential to check websites or phone for details before travelling far.

Protected status | Well protected as State Park and Wilderness, though affected by budget cuts and extensive populations of invasive species.

Opposite: The desert bursts into flower in Anza-Borrego after heavy winter rain in 2005.

There is something particularly special about a desert in full flower. It's partly because this is a rare and exceptional event, only taking place in the wettest of years, but it's also a spectacle of great intrinsic beauty as each flower is displayed against a background of sand or rock, rather than interwined with grass or other foliage. You cannot help but admire the strength and persistence of these fragile plants that may have survived below ground for years waiting for the rains to come.

Deserts characteristically receive less than 250 mm of precipitation a year, but their appearance and vegetation is also strongly affected by the amount of evaporation, the timing of seasonal rainfall, and how frequently wetter years occur. Anza–Borrego is one of the hottest parts of the USA, with temperatures regularly reaching over 40° C, and not uncommonly as high as 50° C, with an average annual rainfall of 175 mm, so it's little surprise that it has all the characteristics of a desert. Some years receive zero rainfall.

To see the best flowers, visit in the wetter years, which normally (though by no means always) coincide with El Niño years, which nowadays are widely predicted. In such years, the flowers are likely to be superb, though it is still worth getting up-to-date information as a hot dry month after rains can wreak havoc with flowering. El Niño years have fallen in 1997–98, 2002–3, 2006–7 and 2009–10, and there is evidence that they are becoming more frequent. In wetter cycles, non-El Niño years can still produce good flowers.

The state park and associated areas of Anza–Borrego are among the best places in the USA to experience the flowering of the desert in an unspoilt environment. The park covers a massive 2,400 km², encompassing a wonderful variety of habitats from just above sea level to over 1,800 m, with a variety of geology to match. In a good early spring (February and March), flowers are everywhere, as far as the eye can see. The beautiful Desert Sand Verbena forms great spreading purple mats, interspersed with lovely clumps of the white-flowered Dune Evening Primrose, yellow fiddlenecks, several species of phacelia in blue or purple, pure white Desert Chicory, and fragrant white spikes of the striking Desert Lily. Other common annuals or herbaceous perennials include the pretty pink cups of Desert Five-spot, orange-yellow Dune Sunflowers, Desert Dandelion, Ghost Flower, monkey flowers and several species of gold-coloured poppy, including the famous California Poppy. Shrubs die back to almost leafless skeletons in drier periods, then burst into leaf and flower after rain. Perhaps the most abundant is the Brittle

Below: The pretty spike of Ghost Flower.

Bottom: A gorgeous clump of Desert Evening Primrose with Sand Verbena beyond.

Opposite: Even the dry stony hilltops can burst into flower after a wet winter.

Bush, which becomes a mass of golden daisy flowers in spring, often growing with the scarlet-flowered Chuparosa, Palo Verde, Desert Willow or the bluish-purple flowered Indigo Bush. One particularly well-adapted shrub is the Ocotillo; its long branches die readily back to bare wood, but produce leaves very quickly whenever there is any rain, doing so as often as 6 times in a year. This strategy allows the bushes to produce their lovely red flowers in almost any year, however dry, making them a crucial supply of nectar for visiting hummingbirds, carpenter bees and other flower feeders.

In recent years, an increasing number and quantity of non-native flowers have begun to appear in the park, such as Tamarisk, African Fountain Grass, and the insidious and abundant Sahara Mustard. These are hard to control, and there seems to be little doubt that they may oust the native species in some circumstances.

Anza–Borrego is worth visiting not just for its spectacular flowers but also for the chance of seeing Desert Bighorn Sheep, desert birds and many other distinctive and specialist desert animals.

Crested Butte area, Colorado Rockies

INFORMATION

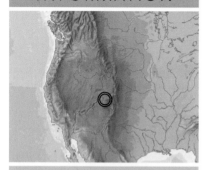

Location | In the central Colorado Rockies, south of Aspen and north of Gunnison, around the town of Crested Butte.

Reasons to go | Fabulous displays of high mountain flowers, with high biodiversity and some endemics, in superb scenery. An excellent annual wildflower festival in July.

Timing | Peak flowers in July and early August, though there will be something of interest anytime between April and October.

Protected status | Well protected by large areas of public ownership, National Forest with designated wilderness.

Opposite: Spectacular alpine flowers including lupines, paintbrushes and Blue Columbine in Rustler's Gulch, near Crested Butte.

There are many wonderful wildflower areas in the Colorado Rockies, but only Crested Butte can claim to be the 'Wildflower Capital of Colorado' as designated under a State Senate resolution passed in 1990. Crested Butte is small and unassuming, but it is surrounded by some wonderful mountains with exceptional displays of flowers. Geologically, it is varied with volcanic, metamorphic and sedimentary rocks, and it owes its origin as a town to the presence of coal and other minerals nearby.

The areas that excel for flowers lie up in the surrounding mountains. Here, winters are exceptionally cold, and 10 m or more of snow may linger until well into July. Summers are warm, but towards late July the 'summer monsoon' begins and thunderstorms or showers drift into the valley, keeping it green and flowery throughout the summer. The most spectacular areas are the mountain meadows, vast open grasslands that lie above or within the highest stretches of woodland, from 3,200 m upwards. These can be reached by road from Kebler Pass, Schofield Pass, Crested Butte Mountain, Lake Irwin and other minor roads north of the town. The flowers here are wonderful, with astonishingly colourful displays of Aspen Sunflowers, blue lupins of several species, red, orange and cream paintbrushes, Blue Larkspur, gentians, several penstemons, gorgeous patches of magenta-pink Parry's Primrose, and the striking blue and white state flower, Blue Columbine in some abundance. It is not uncommon to see hillsides ablaze with flowers as far as the eye can see, such as in Rustler's Gulch or the valley leading from Schofield Park to West Maroon Pass. Yellow-bellied Marmots are abundant in the mountain meadows, and there are plenty of birds and butterflies to be seen.

Higher still, the mountain meadows grade upwards into tundra, which can be full of gems such as the intense blue Sky Pilot, the lovely orange disks of Alpine Sunflower, Arctic Gentian, Dwarf Fireweed, Moss Campion, Mountain Dryad (known as Mountain Avens in Europe) and many others. Lower down, the the deciduous aspen forests can have a rich spring flora including drifts of yellow Glacier Lily, pretty white starflowers, or clumps of the lovely pink Spring Beauty before the canopy closes.

Crested Butte is also home to an annual Wildflower Festival which has grown in its 25-year history to be one of the largest events of its kind, with about 200 flower-related activities from 40 different teachers. It's a good way to discover the flowers of the area but book early (especially if requiring accommodation) or just use the dates as an indication of a good time to come.

174 # San Juan Mountains, Colorado

INFORMATION

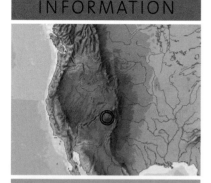

Location | South-west Colorado, lying roughly between the towns of Durango and Montrose, extending east towards Monte Vista.

Reasons to go | Fabulous displays of high mountain flowers, with great biodiversity and some endemics, in superb scenery.

Timing | Peak flowers from late June to early August, though there will be something of interest any time between April and October.

Protected status | Well protected by large areas of public ownership, including wilderness and National Forest, though affected by mining activity and heavy jeep use.

Opposite: Spectacular summer alpine flowers including Rosy Paintbrush at about 3000 metres on Black Bear Pass.

The San Juan Mountains are a beautiful range of jagged, partly volcanic, peaks extending over a wide area of south-west Colorado, reaching almost to the New Mexico border. This is one of those places whose beauty draws you from afar, visible as an intriguing dramatic profile on the skyline from almost any direction. It is a wild and little-known area, with vast stretches of wilderness (particularly the Weminuche Wilderness) and national forests (mainly the San Juan and Uncompahgre National Forests). Unlike the Rockies, the main axis runs roughly east–west, so the mountains catch more precipitation coming up from the south. It is hard to characterize the climate of such a varied area, but in general the winters are very snowy, midsummer is warm and dry, but late summer brings rain in the form of thunderstorms on most days – the so-called summer monsoon.

The flowers are undoubtedly spectacular in many places. Typically, the best displays are at and above the treeline, which here means climbing to 3,300 m or more. These are young mountains, still eroding rapidly, with many areas of scree slopes, cliffs and talus that are interesting but too steep for spectacular displays. In general, the best areas are less steep and have deeper soil, and fortunately there are many such areas. Any good area will have drifts of pink and orange paintbrushes, including the lovely pinkish endemic Hayden's Paintbrush, blue Chiming Bells, golden-yellow Subalpine Arnica, Reddish King's Crown, Blue Larkspurs and monkshood, fleabanes and asters, blue lupins, White Geranium, Jacob's ladders, penstemons and many different yellow composites such as Black-tipped Senecio, to name but a few. Damper areas possess such gems as Fringed Parnassia, magenta Parry's Primrose, white-flowered Marsh Marigold, several louseworts, white orchids and Cottonsedges.

Higher still, some wonderful tundra areas reach up to the jagged peaks. The displays here are usually more open, but often include dramatic displays of Alpine Sunflower, paintbrushes, the more dwarf alpine penstemons, Sky Pilot, Alpine Phlox, Snow Cinquefoil, Alpine Pussytoes, Alpine Arnica and Moss Campion. These are fabulous wild places, far from the sights and sounds of modern life.

Many well-known locations for flowers are reasonably accessible, though 4WD is advisable for most if you are not walking, and be aware that some of the jeep trails can become very congested in busy periods. Yankee Boy Basin, south-west of Ouray at the head of Sneffel's Creek, is probably the best-known location, though it is busy and not that flowery. The higher Governor's Basin, just to the south, is better, and much quieter (though harder to get to).

Below: Lovely masses of Parry's Primrose with Mountain Marsh Marigold high in the mountains at Bullion Lake.

Opposite: The wonderful Alpine Sunflower or Old Man of the Mountains growing close to the snowline in Governor Basin.

Tracks leading westwards from near the summit of the Red Mountain Pass reach some lovely areas, particularly up Porphyry Gulch to Bullion King Lake, or Black Bear Pass. Further south, Ice Lake Basin, up above South Fork Mineral Creek, has a wonderful combination of woodland, meadow and tundra flowers, spectacular scenery, and solitude, as long as you are prepared to walk. Other excellent flower sites include Stony Gulch, on the Silverton side of Stony Pass (about 10 km east of Silverton) which is always good, or the much more famous American Basin, near Cinnamon Pass. This latter suffers a bit from its popularity, but the flowers can be really superb in a good year.

There are many places in the world that might well qualify as some of the 'most flowery'. Many, I suspect, I simply don't know about. Many others, however, I considered and rejected, either because they were not quite flowery enough (possibly because I did not see them at their best), or simply because I did not have time to visit them; or, in a few cases, because I felt they were too inaccessible to be appropriate here. However, the places listed below are wonderful, undoubtedly flowery and worth a visit.

EUROPE

- The Bernese Oberland, Switzerland
- Hintertux, Austria
- The Karawanken Alps, Austria
- The Sar Mountains, Macedonia
- Mount Olympus, Greece

AFRICA

- The Atlas Mountains, Morocco
- The highlands of Ethiopia
- The Drakensberg Mountains, South Africa

ASIA

- Socotra Island, Yemen
- The Lesser Caucasus, Georgia
- The bulb-rich areas of Kyrgyzstan
- Smythe's Valley of Flowers, Zanskar Himalayas, India
- The Sikkim Himalayas, India
- The Hindu Kush, Afganistan

NORTH AMERICA

- The Bruce Peninsula, Ontario, Canada
- Yellow Island Preserve, Puget Sound, Washington State
- Prairie grasslands in Illinois, Dakota, Wisconsin and other parts of North America
- The coastal prairies of the Louisiana–Texas border area
- The Edwards Plateau, Texas
- Big Bend National Park, Texas

CENTRAL AND SOUTH AMERICA

- Baja California, Mexico
- Sierra Madre, and other mountain ranges in Mexico
- Campo Rupestre, Brazil
- The Mediterranean areas of Chile.

EUROPE

Ireland
The Burren
www.burrennationalpark.ie
www.burrenbeo.com

UK
Outer Hebrides
The best contacts are the Uist and Lewis offices of Scottish Natural Heritage.
www.snh.org.uk

For details of the RSPB's Balranald reserve see *www.rspb.org.uk*

The Lizard
The Lizard Countryside Centre, Trelowarren
Tel: +44 (0)1326 221661
www.lizard-peninsula.co.uk
www.naturalengland.org.uk
www.nationaltrust.org.uk

There are visitor information centres at Mullion, Lizard Point and Helston.

Sweden
http://swedentravelnet.com
www.sverigeturism.se/smorgasbord
Both are useful general websites.

www.naturvardsverket.se
Swedish Environmental Protection Agency website, with details of national parks.

Abisko
www.abisko-naturum.nu
www.abisko.nu

Öland
www.stationlinne.se
A good environmental centre on the island.

Estonia
www.keskkonnainfo.ee
The Estonian Environmental Information Centre.

France
Information centres of variable quality, some excellent, are found in most of the main towns and villages in the national parks.

Vercors
www.parc-du-vercors.fr

Ecrins
www.ecrins-parcnational.fr

Cevennes and Causses
www.parc-grands-causses.fr
www.cevennes-parcnational.fr

Central pyrenees
www.parc-pyrenees.com

Spain
Ordesa and Monte Perdido
www.ordesa.net/ordesa
There is a visitor centre at Ordesa and general information centres at Torla, Bielsa and at the Valle de Pineta Parador.

Picos de Europa
http://reddeparquesnacionales.mma.es/parques/picos/index.htm
The official national park website, somewhat difficult to navigate.

There are visitor centres in Posada de Valdeon, Cangas de Onis, and Buferrera.

Sierra de Grazalema
www.grazalemaguide.com
A really useful website with lots of information.
Permits can be obtained from the park information centre in Cadiz.

There is also an information centre in Zahara de La Sierra. For details of the park in Spanish see *www.cadiz-turismo.com/parquesnaturales/sierradegrazalema/sierradegrazalema.php*

Portugal
The Alentejo and the Costa Vicentina Natural Park Head Office
Rua Serpa Pinto nĚ 32, 7630-174 Odemira
Tel: +351 283 322 735

Switzerland
www.myswitzerland.com
A search on 'Engadin' will provide lots of useful information about the area.

Italy
Dolomites
www.dolomitipark.it
www.val-gardena.com
www.dolomiti.org

Lake Garda
www.parks.it/parco.alto.garda.bresciano

Piano Grande
www.sibillini.net.

Gargano
www.parcogargano.it

Slovenia

www.tnp.si/national_park
The site for Triglav National Park, including contact details for the excellent information centres in Trenta and Bled.

www.bohinj.si/alpskocvetje/eng/o_festivalu.php
Details of the wildflower festival.

Romania

www.romaniatourism.com
General information and a section on national parks.

www.fundatia-adept.org
www.discovertarnavamare.org
Two sites run by Fundatia ADEPT, a charity supporting conservation and economic regeneration of the Transylvanian grasslands.

Greece

www.eepf.gr
Hellenic Society for the Protection of Nature (content in Greek only).

www.gnto.gr
A wealth of general information from the Greek National Tourism Organization.

Turkey

www.goturkey.com
General tourist information including some details about national parks.

Cyprus

www.european-foresters.org/CyprusGCM/Troodos
What to see and do in Troodos National Forest Park.

www.visitcyprus.com
General information and Akamas Nature Trail guides.

AFRICA

Tanzania

http:// tanzaniaparks.com

South Africa

www.namaqualand.com
General information about Namaqualand.

www.kamieskroonhotel.com
Kamieskroon Hotel, for accommodation, courses and information.

Goegap Nature Reserve

www.sa-venues.com/game-reserves/nc_goegap.htm

Richtersveld National Park
www.sanparks.org/parks/richtersveld

Namaqua National Park
www.sanparks.org/parks/namaqua .

Niewoudtville
www.nieuwoudtville.com

Hantam Botanical Garden
www.sanbi.org
Follow links to Hantam.

Table Mountain National Park
www.sanparks.org/parks/table_mountain
Includes contact details for BuffelsfonteinVisitor Centre

Kogelberg Nature Reserve
www.capenature.org.za/reserves.htm
Tel: +27 028 271 5138, e-mail admin@kogelbergbiosphere.co.za

Fernkloof Nature Reserve
http://fernkloof.com

ASIA

Georgia

www.cac-biodiversity.org/geo
Information is very limited but this site focussed on plant genetic resources provides some useful links.

Iran

Travelling in Iran is difficult as an individual, especially if you do not speak Farsi. The best way to visit is as part of a tour.

Kazakhstan

www.aksuinn.com
This excellent guesthouse in Dzhabagly can organize travel in the area and provide guides for the reserve.

China

Trips need to be planned with some care. Within China, the China International Travel Service (CYTS) and China Travel Service (CTS) can arrange tours. Outside China various firms run natural history tours led by experts to the region.

AUSTRALASIA

Australia

www.wildflowerswa.com
Information about visitors centres, accommodation, wildflower walks, and flowering times.

www.stirlingrange.com.au
The Stirling Range Retreat (+61 (0)8 9827 9229), run by botanists and offering information, guided walks and tours, and accommodation, though it is up for sale at the time of writing.

New Zealand

www.doc.govt.nz/parks-and-recreation/national-parks/fiordland
Information about Fiordland from the Department of Conservation.

www.fiordland.org.nz
Useful travel information.

NORTH AMERICA

Canada

www.pc.gc.ca
The Parks Canada site is a useful starting point, though not good on flowers. There is a visitor centre at Waterton.

The Glacier National Park (USA) website is www.nps.gov/glac/index.htm

USA

www.nps.gov
The best starting point for national parks in the US.

Olympic National Park

Excellent visitor centres at Hurricane Ridge (seasonal opening), Hoh, Forks, and the main year-round visitor centre at Port Angeles.

Kalmath-Siskiyou

www.fs.fed.us/wildflowers/communities/serpentines/center
Information from the US Forest Service.

There are visitor centres in Ashland, Mount Shasta, Medford, and just north of Galice at the Smullin.

Carrizo Plain

www.blm.gov/ca/st/en/fo/bakersfield/Programs/carrizo.html
This site from the Bureau of Land Management has links to other sites.

The excellent Goodwin Education Centre is open from early December to late May, Thursday to Sunday, 9:00 am to 4:00 pm.

Tehatchapi

www.tejonconservancy.org
Tours within the Tejon Ranch area.

www.theodorepayne.org
The Theodore Payne Foundation runs a hotline and website (covering most of southern California) in season.

www.parks.ca.gov/?page_id=627
Details of the Antelope Valley Poppy Reserve from California State Parks, including hotline numbers.

Anza-borrego

www.parks.ca.gov/?page_id=638
www.desertusa.com
Both sites have wildflower updates.

Crested Butte

www.crestedbuttewildflowerfestival.
www.crestedbutte-co.gov

An excellent book on the area by Katherine Darrow (see bibliography) not only helps in the identification of most flowers of the area, but also suggests various places to see flowers within easy reach of Crested Butte.

San Juan Mountains

www.fs.fed.us/r2/sanjuan
Information from the US Forest Service, including contact details for the Silverton Public Lands Center, which makes an excellent starting poin for touring the area.

www.ouraycolorado.com
General information about the Ouray area.

Most of the companies that run a wide range of botanical tours are UK based, and they attract participants from around the world. In other countries, there are often foreign tours arranged through University Adult Education schemes or Botanical societies, though there are more limited in scope.

Ace Cultural Tours

acestudytours.co.uk   *+44(0)1223 835 055*

A few botanical tours.

Advantour

www.advantour.com

Local tour operators specializing in Silk Road countries.

Alpine Garden Society

www.alpinegardensociety.net/tours   *+44 (0)1386 554 790 (UK)*

Organize tours to flower-rich mountain areas and also operate tours in conjunction with Greentours, (see below).

ATG Oxford

www.atg-oxford.co.uk   *+44 (0)1865 315 678*

A limited range of tours that include flowers.

Botanical Expeditions

http://botanicalexpeditions.com   *(800) 252-4910 or (408) 252-4910*

A division of Betchart Expeditions, based in California. Run a number of botanical tours.

Caravan Sahra

www.caravansahra.com

A general tour operator based in Tehran with experience of many botanical tours.

Estonian Nature Tours

www.naturetours.ee   *+372 477 8214*

Botanical and other tours.

Greentours

www.greentours.co.uk   *+44 (0)1298 83563*

A wide range of botanical and general natural history tours.

Iberian Wildlife Tours

www.iberianwildlife.com

Offer several botanical tours to flower-rich areas of Spain and Portugal.

Journey Latin America

www.journeylatinamerica.co.uk   *+44 (0)20 8747 8315*

A limited range of flower tours in South America.

Natural History Travel

Contact Dr Bob Gibbons on *bobgibbons@btinternet.com*

A small programme of botanical and general natural history tours.

Naturetrek

www.naturetrek.co.uk   *+44 (0)1962 733051*

A huge range of tours, including flower-related trips.

Nature Quest New Zealand

www.naturequest.co.nz   *+64 3 489 8444*

Run tours in the Fiordland National Park area.

The Travelling Naturalist

www.naturalist.co.uk   *+44(0)1305 267994*

Mainly bird-orientated, but some tours have a strong botanical bias.

Wildlife Travel

www.wildlife-travel.co.uk   *+44(0)1954 713575*

A good range of botanical and natural history tours.

Akeroyd, John. *The historic countryside of the Saxon Villages of southern Transylvania*. Fundatia ADEPT, Saschiz, Romania. 2006.

Biek, David. *Flora of Mount Rainier National Park*. Oregon State University Press, Corvallis. 2000.

Darrow, Katherine. *Wild About Wildflowers. Extreme botanising in Crested Butte*. Heel and Toe Publishing, Fort Collins, Colorado. 2006.

Faber, Phyllis. *California's Wild Gardens*. University of California Press, Berkeley. 2005.

Farino, Teresa & Lockwood, Mike. *Traveller's Nature Guide: Spain*. OUP, Oxford. 2003.

Gibbons, Bob. *Traveller's Nature Guide: Greece*. OUP, Oxford. 2003.

Gibbons, Bob. *Traveller's Nature Guide: France*. OUP, Oxford. 2003.

Gibbons, Bob. *Wild France*. New Holland Publishers, UK. 2009.

Gurche, Charles. *Washington's Best Wildflower Hikes*. Westcliffe Publishers, Colorado. 2004.

Holubec, V & Krivka, P. *The Caucasus and its Flowers*. Loxia, Prague. 2006.

Irwin, Pamela & David. *Colorado's Best Wildflower Hikes 3: San Juan Mountains*. Westcliffe Publishers, Colorado. 2006.

Jepson, Tim. *Wild Italy*. Sheldrake Press, London. 1994.

Joyce, Peter. *Flower Watching in the Cape*. Struik. Cape Town. 2004.

Manning, John. *Field Guide to Fynbos*. Struik. Cape Town. 2007.

Nevill, Simon. *Traveller's Guide to the Parks and Reserves of Western Australia*. 2001. Simon Nevill Publications, Fremantle, Australia.

Richards, John. *Mountain Flower Walks: The Greek Mainland*. Alpine Garden Society, Pershore. 2008.

Salmon, John. *A Field Guide to the Alpine Plants of New Zealand*. Godwit Publishing, Auckland. 1992.

Sterner, Rikard. *Ölands Kärlväxtflora*. Förlagstjänsten, Stockholm.1986. (A flora of Öland with a section in English).

Stewart, Jon Mark. *Mojave Desert Wildflowers*. Jon Stewart Photography, Albuquerque. 1998.

African Fountain Grass *Pennisetum setaceum*

Akamas Alison *Alyssum akamassicum*

Akamas Knapweed *Centaurea akamantis*

alkanets *Alkanna* spp.

Alpine Arnica *Arnica alpina*

Alpine Aster *Aster alpinus*

Alpine Bartsia *Bartsia alpina*

Alpine Cushion *Donatia novae-zelandiae*

Alpine Forget-me-not *Myosotis alpestris*

Alpine Laburnum *Laburnum alpinum*

Alpine Marsh Orchid *Dactylorhiza alpestris*

Alpine Pasque Flower *Pulsatilla alpina*

Alpine Phlox *Phlox condensata*

Alpine Pussytoes *Antennaria alpina*

Alpine Rose *Rosa pendulina*

Alpine Snowbell *Soldanella alpina*

Alpine Squill *Scilla bifolia*

Alpine Sunflower or Old Man of the Mountains *Hymenoxys grandiflora*

Alpine Woundwort *Stachys alpina*

Annual Androsace *Androsace maxima*

Apennine Anemone *Anemone apennina*

Aphrodite's Knapweed *Centaurea veneris*

Aphrodite's Spurge *Euphorbia veneris*

Aphyllanthes *Aphyllanthes monspeliensis*

Arctic Bellflower *Campanula uniflora*

Arctic Gentian *Gentiana algida*

Arctic Orchid *Platanthera oligantha*

Arctic Rhododendron *Rhododendron lapponicum*

Arnica *Arnica alpina*

Ashy Cranesbill *Geranium cinereum*

Aspen Sunflowers *Helianthella quinquenervis*

Aubretia *Aubrieta columnae*

Avalanche Lily *Erythronium montanum*

Aymonin's Orchid *Ophrys aymoninii*

Balsamroot *Balsamorhiza sagittata*

Basil Thyme *Acinos arvensis*

Bastard Balm *Melittis melissophyllum*

Bear Grass *Xerophyllum tenax*

Bear's Ear Primula *Primula auricula*

beardtongues *Penstemon* spp.

Beargrass *Xerophyllum tenax*

Bee Orchid *Ophrys apifera*

Beetle Daisy *Gorteria diffusa*

Bell Heather *Erica cinerea*

Bell-fruited Mallee *Eucalyptus preissiana*

Bentham Lupine *Lupinus benthamii*

Bigroot Cranesbill *Geranium macrorrhizum*

Bird's Eye Primrose *Primula farinosa*

Bird's Foot Trefoil *Lotus corniculatus*

Bird's Nest Orchid *Neottia nidus-avis*

Bistort *Polygonum bistorta*

Black Pea *Thermopsis barbata*

Black Pine *Pinus nigra* var. *pallasiana*

Black-tipped Senecio *Senecio atratus*

Bloody Cranesbill *Geranium sanguineum*

Blue Beardtongue *Penstemon albertinus*

Blue Bugle *Ajuga genevensis*

Blue Clematis *Clematis columbiana*

Blue Columbine *Aquilegia caerulea*

Blue Larkspur *Delphinium barbeyi*

Bluebell *Hyacinthoides non-scripta*

Bluish Paederota *Paederota bonarota*

Bog Asphodel *Narthecium ossifragum*

Bog Rosemary *Andromeda polifolia*

Bolander's Onion *Allium bolanderi*

Bornmueller's Ophrys *Ophrys bornmuelleri*

Box *Buxus sempervirens*

Brittle Bush *Encelia farinosa*

Broad-leaf Arnica *Arnica latifolia*

Broadleaf Lupine *Lupinus latifolius*

Broad-leaved Snowdrop *Galanthus elwesii*

Brown-eyed Susan *Gaillardia aristata*

Bulley's Rock Jasmine *Androsace bulleyana*

Bunchberry Cornus *canadensis* or *C. unalaschkensis* (recently separated)

Burning Bush *Dictamnus albus*

Burnt Candytuft *Aethionema saxatile*

Burnt Orchid *Orchis ustulata*

Bushman's Candle *Sarcocaulon crassicaule*

Butterfly Lily *Wachendorfia* spp.

Calabrian Soapwort *Saponaria calabrica*

California Poppy *Eschscholzia californica*

Carmel Ophrys *Ophrys attica*

Carnic Lily *Lilium carniolicum*

Cascade Azalea *Rhododendron albiflorum*

Cat-tails *Bulbinella* spp.

Caucasian Anemone *Anemone caucasica*

Caucasian Rhododendron *Rhododendron caucasicum*

Caucasian Snake's Head *Fritillaria latifolia*

Cedar of Lebanon *Cedrus libani*

Ceratocephalus *Ceratocephalus falcatus*

Chalk Milkwort *Polygala calcarea*

Champagne Orchid *Orchis champagneuxii*

Chickweed Wintergreen *Trientalis europaea*

Chiming Bells *Mertensia ciliata*

Christmas Rose *Helleborus niger*

Chuparosa *Beloporone californica*

Cilician Fir *Abies cilicica*

Cliff Indian Paintbrush *Castilleja rupicola*

Cliff Penstemon *Penstemon rupicola*

Cliff Primrose *Douglasia laevigata*

Clustered Bellflower *Campanula glomerata*

Coiled-beak Lousewort, red form *Pedicularis bracteosa* var. *atrosanguinea*

Columbia Lily *Lilium columbianum*

columbines *Aquilegia* spp.

Common Broom *Cytisus scoparius*

Common Bugle *Ajuga reptans*

Common Lungwort *Pulmonaria officinalis*

Common Milkwort *Polygala vulgaris*

Common Restharrow *Ononis repens*

Common Rock Rose *Helianthemum nummularium*

Common Spotted Orchid *Dactylorhiza fuchsii*

Common Toadflax *Linaria vulgaris*

Common Vegetable Sheep *Raoulia eximia*

Comper's Orchid *Comperia comperiana*

Coral-root Bittercress *Cardamine bulbifera*

Corky Honey-myrtle *Melaleuca suberosa*

Corn Cockle *Agrostemma githago*

Corn Marigold *Chrysanthemum segetum*

Cornflower *Centaurea cyanus*

Cornish Heath *Erica vagans*

cottonheads *Conostylis* spp.

Cowberrys *Vaccinium vitis-idaea*

Cowslip *Primula veris*

cow-wheats *Melampyrum* spp.

Creeping Avens *Geum reptans*

Creeping Azalea *Loiseleuria procumbens*

Creeping Globularia *Globularia repens*

Cretan Eryngo *Eryngium creticum*

Crown Anemone *Anemone coronaria*

Crown Imperial *Fritillaria imperialis*

Crown Vetch *Coronilla varia*

Cypress Spurge *Euphorbia cyparrisias*

Cypriot Tulip *Tulipa cypria*

Cyprus Broomrape *Orobanche cypria*

Cyprus Cedar *Cedrus brevifolia*
Cyprus Crocus *Crocus cyprius*
Dark Columbine *Aquilegia atrata*
Davidson's Penstemon *Penstemon davidsonii* var. *menziesii*
Dense-flowered Orchid *Orchis intacta*
Desert Candle *Caulanthus inflatus*
Desert Chicory *Rafinesquia neomexicana*
Desert Dandelion *Malacothrix glabrata*
Desert Five-spot *Eremalche rotundifolia*
Desert Lily *Hesperocallis undulata*
Desert Primrose *Grielum humifusum*
Desert Sand Verbena *Abronia villosa*
Desert Willow *Chilopsis linearis*
Devil's Claw *Physoplexis comosa*
Diamond Flower (also Violet Cress) *Ionopsidium acaule*
Diapensia *Diapensia lapponica*
Dittany *Origanum dictamnus*
Douglas Fir *Pseudotsuga menziesii*
Drooping Bittercress *Cardamine enneaphyllos*
Dropwort *Filipendula vulgaris*
Drumsticks *Zaluzianskya violacea*
Dune Evening Primrose *Oenothera deltoides*
Dune Sunflowers *Geraea canescens*
Dusky Cranesbill *Geranium phaeum*
Dutchman's Pipe *Aristolochia* spp.
Dwarf Alpenrose *Rhodothamnus chamaecistus*
Dwarf Cornel *Cornus suecica*
Dwarf Fireweed *Chamaerion latifolia*
Dwarf Goldenrod *Solidago simplex* ssp. *simplex* var. *nana*
Dwarf Lupin *Lupinus lepidus* (*L. lyalii*)
Dwarf Mouse-ear *Cerastium pumilum*
Dyer's Alkanet *Alkanna lehmannii*
Dyer's Greenweed *Genista tinctoria*
Early Marsh Orchid *Dactylorhiza incarnata*
Early Purple Orchid *Orchis mascula*
Edelweiss *Leucogenes grandiceps*
Elder-flowered Orchid *Dactylorhiza sambucina*

Elegant Ophrys *Ophrys elegans*
English Iris *Iris latifolia*
Entire-leaved Clematis *Clematis integrifolia*
Evening Snow *Linanthus dichotomus*
everlastings *Schoenia* and *Rhodanthe* spp.
eyebrights *Euphrasia* spp.
False Everlasting *Phaenocoma prolifera*
False Sainfoin *Vicia onobrychioides*
Fan-leaf Cinquefoil *Potentilla flabellifolia*
Fan-lipped Orchid *Orchis collina* (*saccata*)
feather flowers *Verticordia* spp.
fenugreeks *Trigonella* spp.
Few-flowered Orchid *Orchis pauciflora*
fiddlenecks *Amsinckia* spp.
Field Poppy *Papaver rhoeas*
Five-leaved Bittercress *Cardamine pentaphylla*
Flett's violet *Viola flettii*
Fly Orchid *Ophrys insectifera*
Foam Flower *Tiarella unifoliata*
Four o'clocks or Common Parakeelya *Calandrinia polyandra*
Four-spotted Orchid *Orchis quadripunctata*
foxtail lilies *Eremurus* spp.
Fragrant Orchid *Gymnadenia conopsea*
Frog Orchid *Coeloglossum viride*
Fringed Parnassia *Parnassia fimbriata*
Garden Angelica *Angelica archangelica*
Gargano Deadnettle *Lamium garganicum*
German Asphodel *Tofieldia calyculata*
Ghost Flower *Mohavea confertiflora*
Ghost Orchid *Epipogium aphyllum*
Giant Andersonia *Andersonia echinocephala*
Giant Fennel *Ferula communis*
Giant or Greater Snowdrop *Galanthus elwesii*
Giant Orchid *Barlia robertiana*
Giant Vegetable Sheep *Haastia pulvinaris*
Glacier Crowfoot *Ranunculus glacialis*
Glacier Lily *Erythronium Erythronium grandiflorum*

Glacier Lily *Erythronium grandiflorum*
Globe Flowers *Trollius europaeus*
Globe-flowered Coronilla *Coronilla globosa* or *Securigera globosa*
Globe-flowered Orchids *Traunsteinera globosa*
Glossy-eyed Parachute Daisy *Ursinia cakilefolia*
Golden Daisy *Erigeron aureus*
Golden Drop *Onosma* spp.
Golden Spaniard *Aciphylla* aurea
Golden Waitzia *Waitzia nitida*
Goldenrod *Solidago virgaurea*
Goldfields *Lasthenia minor*
Gouan's Buttercup *Ranunculus gouanii*
grape hyacinths *Muscari* spp.
Grass of Parnassus *Parnassia palustris*
Great Milkwort *Polygala major*
Great Yellow Bumblebee *Bombus distinguendus*
Greater Butterfly Orchid *Platanthera chlorantha*
Greater Yellow Rattle (Asturian subspecies) *Rhinanthus serotinus* ssp. *asturicus*
Greek Fritillary *Fritillaria graeca*
Greek Spiny Spurge *Euphorbia acanthothamnos*
Green-winged Orchid *Orchis morio*
Greig's Tulip *Tulipa greigii*
guinea flowers *Hibbertia* spp.
gum cistus *Cistus ladanifer*
Hairy Greenweed *Genista pilosa*
Halfmens Tree *Pachypodium namaquanum*
Haller's Pasque Flower *Pulsatilla halleri*
Harebell *Wahlenbergia albomarginata*
Harsh Hakea *Hakea prostrata*
Hayden's Paintbrush *Castilleja haydenii*
Hay Rattle *Rhinanthus* spp.
Heart-flowered Serapias *Serapias cordigera*
Heartsease *Viola tricolor*
Heath Spotted Orchid *Dactylorhiza maculata* ssp. *ericetorum*
Hebridean Marsh Orchid *Dactylorhiza ebudensis*
Hebridean Spotted Orchid *Dactylorhiza hebridensis*
Hedgehog Broom *Erinacea anthyllis*
Heldreich's Anemone *Anemone heldreichii*

Hellebore *Helleborus orientalis*
Henderson's Horkelia *Horkelia hendersonii*
Herb Paris *Paris quadrifolia*
Herbaceous Periwinkle *Vinca herbacea*
Hillside Daisy *Monolopia lanceolata*
Hoary Plantain *Plantago media*
Hoary Rock Rose *Helianthemum canum*
Holy Orchid *Orchis sancta*
honey-myrtle *Melaleuca* spp.
Horned Orchid *Ophrys oestrifera*
Horned Pansy *Viola cornuta*
Horseshoe Orchid *Ophrys ferrum-equinum*
Horseshoe Vetch *Hippocrepis comosa*
Indian Head-dress *Tracyina rostrata*
Indigo Bush *Psorothamnus* spp.
Intermediate Periwinkle *Vinca difformis*
Israeli Ophrys *Ophrys israelitica*
Italian Orchid *Orchis italica*
Jacob's ladders *Polemonium* spp.
Jarrah *Eucalyptus* spp.
Jeffrey's Shooting Star *Dodecatheon jeffreyi*
Joint Pine *Ephedra campylopoda*
Joshua Tree *Yucca brevifolia*
Jove's Flower *Lychnis flos-jovis*
Judas Tree *Cercis siliquastrum*
Julian Poppy *Papaver alpinum* ssp. *ernesti-mayeri*
Kalmiopsis *Kalmiopsis leachiana*
Karatau Onion *Allium karataviense*
Karoo Violet *Aptosimum dichotomum*
Kerry Lily *Simethis planifolia*
Kidney Vetch *Anthyllis vulneraria* ssp. *vulneraria*
King's Crown *Rhodiola integrifolia* or *Sedum integrifolium*
King-of-the-Alps *Eritrichium nanum*
Kipunji *Rungwecebus kipunji*
Kotschy's Orchid *Ophrys kotschyi*
Kykko Buttercup *Ranunculus kykkoensis*
Lacy Phacelia *Phacelia tanacetifolia*
Ladies Bedstraw *Galium verum*
Lady's Bedstraw *Galium verum*
Lady Loch's Glory of the Snows *Chionodoxa (Scilla) lochii*
Lady Orchid *Orchis. purpurea* =?
Lady's Slipper Orchid *Cypripedium calceolus*

Lambertia *Lambertia uniflora*

Lange's Orchid *Orchis langei*

Lapethos Ophrys *Ophrys lapethica*

Large Blue Alkanet *Anchusa italica*

Large Mountain Monkey Flower *Mimulus tilingii* var. *caespitosus*

Large Speedwell *Veronica austriaca*

Large Venus' Looking-glass *Legousia speculum-veneris*

Large-flowered Butterwort *Pinguicula grandiflora*

Large-flowered Campion *Silene elisabethae*

Large-flowered Mat Daisy *Raoulia grandiflora*

Large-flowered Meadow-Rue *Thalictrum aquilegifolium*

Least Primula *Primula minima*

Leontice *Leontice leontopetalum*

Leopard's Bane *Doronicum pardalianches*

Lesser Butterfly Orchid *Platanthera bifolia*

Lesser Solomon's Seal *Polygonatum odoratum*

Lewis's Monkey-flower *Mimulus lewisii*

Lily-of-the-Valley *Convallaria majalis*

Ling *Calluna vulgaris*

Lizard Orchid *Himantoglossum hircinum*

Lobel's False Helleborine *Veratrum lobelianum*

Long-leaved Butterwort *Pinguicula longifolia*

louseworts *Pedicularis* spp.

lupins *Lupinus* spp.

Magenta Paintbrush *Castilleja parviflora*

Maiden Pink *Dianthus deltoides*

Marsh Pagoda *Mimetes hirtus*

Marsh Rose *Orothamnus zeyheri*

Marsh Valerian *Valeriana dioica*

Martagon Lily *Lilium martagon*

Masterwort *Astrantia major*

Mat Daisy *Celmisia* spp.

mayweeds *Anthemis* spp.

Meadow Clary *Salvia pratensis*

Meadow Saffron *Colchicum autumnale*

Meadow Saxifrage *Saxifraga granulata*

Mediterranean Catchfly or Pink Pirouette *Silene colorata*

Military Orchid *Orchis militaris*

Mirror Orchid *Ophrys speculum*

monkey flowers *Mimulus* spp.

Monte Baldo Anemone *Anemone baldensis*

Moonwort *Botrychium lunaria*

Mount Ashland Lupine *Lupinus aridus* var. *ashlandensis*

Mount Cook Lily *Ranunculus lyalii*

Mountain Avens *Dryas octopetala*

mountain bells *Darwinia* spp.

Mountain Dryad *Dryas octopetala*

Mountain Kidney Vetch *Anthyllis montana*

Mountain Kunzea *Kunzea montana*

Mountain Lady's Slipper Orchid *Cypripedium montanum*

Mountain Locoweed *Oxytropis monticola*

Mountain Marsh Marigold *Caltha leptosepala*

Mountain Owl Clover *Orthocarpus imbricatus*

Mountain Pansy *Viola altaica*

Mountain Sorrel *Oxyria digyna*

Mountain-heath *Sphenotoma squarrosa*

mulleins *Verbascum* spp.

Musk Thistle *Carduus nutans*

Namaqua Parachute Daisy *Ursinia cakilefolia*

Namaqualand Daisy *Dimorphoteca sinuata*

Narrow-leaved Milk-vetch *Astragalus angustifolia*

Narrow-leaved Vetch *Vicia tenuifolia*

Needle-leaved Chorizema *Chorizema aciculare*

Nodding Sage *Salvia nutans*

Northern Golden Rod *Solidago multiradiata* var. *scopulorum*

Northern Wolfsbane *Aconitum septentrionale*

Norwegian Sandwort *Arenaria norvegica* ssp. *norvegica*

Norwegian Wintergreen *Pyrola norvegica*

Ocotillo *Fouquieria splendens*

Öland Rock Rose *Helianthemum oelandicum* ssp. *oelandicum*

Olympic Larkspur *Delphinium glareosum*

Olympic Onion *Allium crenulatum*

One-flowered Wintergreen *Moneses uniflora*

Orange Lily *Lilium bulbiferum*

Orange Mullein *Verbascum phlomoides*

Oregon Bensoniella *Bensoniella oregana*

Oriental Beech *Fagus orientalis*

Oriental Bugle *Ajuga orientalis*

Owl's Clover *Castilleja* spp. (formerly *Orthocarpus* spp.)

Ox-eye Daisy *Leucanthemum vulgare*

Oxlip *Primula elatior*

paintbrushes *Castilleja* spp.

Pale Dog Violets *Viola lactea*

Palo Verde *Cercidium floridum*

Paper-white Narcissus *Narcissus paptraceus*

Parry's Mallow *Eremalche parryi*

Parry's Primrose *Primula parryi*

Partridge Foot *Luetkea pectinata*

Pasque Flower *Pulsatilla vulgaris*

Peacock Anemones *Anemone pavonina*

Penwiper Plant *Notothlaspi rosulatum*

Perennial Cornflower *Centaurea montana*

Persian Cyclamen *Cyclamen persicum*

Pheasant's Eye *Adonis annua*

Pheasant's Eye Narcissus *Narcissus poeticus*

pincushions *Chaenactis* spp. (USA)

pincushions *Leucospermum* spp. (South Africa)

Pink Butterfly Orchid *Orchis papilionacea*

Pink Flax *Linum suffruticosum*

Pink Hawksbeard *Crepis rubra*

Plane-leaved Buttercup *Ranunculus platanifolius*

Pink Sandwort *Arenaria purpurascens*

Pink Velleia *Velleia rosea*

Piper's Bellflower *Campanula piperi*

Plutonian Chamomile *Anthemis plutonia*

Pompom Head *Cephalipterum drummondii*

Prickly Moses *Acacia pulchella*

Pride of Table Mountain *Disa uniflora*

Prince of Wales Heath *Erica perspicua*

Prince's Pine (Umbellate chickweed) *Chimaphila umbellata*

Prostrate Plum *Prunus prostrata*

Psycrophila *Psycrophila obtusa*

Purple Bugloss *Echium plantagineum*

Purple Oxlip *Primula amoena*

Purple Phlomis *Phlomis purpurea*

Purple Rock-cress *Arabis purpurea*

Purple Saxifrage *Saxifraga oppositifolia*

Pyramidal Bugle *Ajuga pyramidalis*

Pyramidal Orchid *Anacamptis pyramidalis*

Pyrenean Avens *Geum pyrenaicum*

Pyrenean Fritillary *Fritillaria pyrenaica*

Pyrenean Poppy *Papaver lapeyrousianum*

Pyrenean Ramonda *Ramonda pyrenaica*

Pyrenean Saxifrage *Saxifraga longifolia*

Pyrenean Squill *Scilla liliohyacinthus*

Pyrenean Whitlow-grass *Petrocallis pyrenaica*

Queen of Sheba Orchid *Thelymitra variegata*

Queen's Cup *Clintonia uniflora*

Quiver Tree *Aloe dichotoma*

Ragged Robin *Lychnis flos-cuculi*

rampions *Phyteuma* spp.

Red and White Baneberry *Actaea rubra*

Red Campion *Silene dioica*

Red Clover *Trifolium pratense*

Red Heather *Phyllodoce empetriformis*

Red Mountain Heather *Phyllodoce empetriformis*

Red Rock Cushion *Phyllachne rubra*

Red Valerian *Centranthus ruber*

Red Viper's Bugloss *Echium russicum*

Red-berried Mistletoe *Viscum cruciatum*

Reinhold's Orchid *Ophrys reinholdii*

Rhaetian Poppy *Papaver rhaetica*

Rhinoceros Bush or Renosterbos *Elytropappus rhinocerotis*

Rock Cushion *Phyllachne colensoi*

Rock Daphne *Daphne petraea*

Rock Soapwort *Saponaria ocymoides*

rock jasmines *Androsace* spp.

Rocky Mountain Dwarf Primrose *Douglasia montana*

Roman Marsh Orchid *Dactylorhiza romana*

Rosemary *Rosmarinus officinalis*

Rose-root *Rhodiola rosea* = *Sedum roseum*

Rough-leaved Alumroot *Heuchera cylindrica*

Round-headed Rampion *Phyteuma orbiculare*

Round-leaved Restharrow *Ononis rotundifolia*

Sahara Mustard *Brassica tournefortii*

Sainfoin *Onobrychis viciifolia*

San Joaquin Blazing Star *Mentzelia pectinata*

sand crocuses *Romulea* spp.

Scallop Flower *Hakea cucullata*

Sea Campion *Silene uniflora*

Sea Holly *Eryngium. maritimum*

Sea Medick *Medicago marina*

Seguier's Buttercup *Ranunculus seguieri*

Self-heal *Prunella vulgaris*

Sheep's Bit *Jasione* spp.

shooting stars *Dodecatheon* spp.

Shore Campion *Silene littorea*

Showy Locoweed *Oxytropis splendens*

Shrubby Buckler-mustard *Biscutella frutescens*

Shrubby Cinquefoil *Dasiphora fruticosa = Potentilla fruticosa*

Silver Fir *Abies amabilis*

Silver Tree *Leucadendron argenteum*

Siskiyou Fritillary *Fritillaria glauca*

Sky Pilot *Polemonium viscosum*

Small Pasque Flower *Pulsatilla pratensis*

Small Spider Orchid *Ophrys araneola*

Small Yellow Foxglove *Digitalis lutea*

Small White Orchid *Leucorchis albida*

Small-flowered Gorse *Ulex parviflorus*

Small-flowered Penstemon *Penstemon procerus*

Snow Buttercup *Ranunculus eschscholtzii*

Snow Cinquefoil *Potentilla nivea*

Snow Gentian *Gentiana nivalis*

snowbells *Soldanella* spp.

snowdrops *Galanthus* spp.

Snowdrop Windflower *Anemone sylvestris*

Snow-in-Summer *Cerastium candidissimum*

Snowy Mespil *Amelanchier ovalis*

Sorrels *Oxalis* spp.

Southern Bartsia *Parentucellia latifolia*

Southern Greenweed *Genista radiata*

Spanish Bluebell *Hyacinthoides hispanica*

Spanish Fir *Abies pinsapo*

Spanish Fritillary *Fritillaria hispanicus*

Spanish Iris *Iris xiphium*

Sparrow's Egg Lady's Slipper Orchid *Cypripedium passerinum*

spiderheads *Serruria* spp.

Spiny Broom *Genista acanthoclada*

Spiny Knapweed *Centaurea spinosa*

Spotted Saxifrage *Saxifraga bronchialis*

Spreading Phlox *Phlox diffusa*

Spring Beauty *Claytonia lanceolata*

Spring Gentian *Gentiana verna*

Spring Pasque Flower *Pulsatilla vernalis*

Spring Squill *Scilla verna*

Spring Vetch *Lathyrus vernus*

Spruner's Orchid *Ophrys spruneri*

St Bruno's Lily *Paradisea liliastrum*

St Vincent Squill *Scilla vicentina*

St. Bernard's Lily *Anthericum liliago*

St. Bruno's Lily *Paradisea liliastrum*

St. Dabeoc's Heath *Daboecia cantabrica*

starflowers *Calytrix* spp. (Australia)

starflowers *Smilacina stellata* (USA)

stars-of-Bethlehem *Ornithogalum* spp.

Stellera *Stellera chamaejasme*

Sticky Purple Geranium *Geranium viscosissimum*

Stinking Juniper *Juniperus foetidissima*

Stirling Range Bottlebrush *Beaufortia heterophylla*

Stirling Range Coneflower *Isopogon latifolius*

Striped Coral-root *Corallorhiza striata*

Subalpine Arnica *Arnica rydbergii*

Subalpine Fir *Abies lasiocarpa*

sundews *Drosera* spp.

sunflaxes *Heliophila* spp.

Swainson's Woundwort *Stachys swainsonii*

Swallow-wort *Vincetoxicum hirundinaria*

Sword-leaved Helleborine *Cephelanthera longifolia*

Syrian Orchid *Orchis syriaca*

Tassel Hyacinth *Muscari comosum*

Thimbleberry *Rubus parviflorus*

Three-cornered Leek *Allium triquetrum*

Three-leaved Colchicum *Colchicum triphyllum*

Three-leaved Snowflake *Leucojum trichophyllum*

Three-Spot Mariposa-Lily *Calochortus apiculatus*

Thrift *Armeria maritima*

Thyme Broomrape *Orobanche alba*

tick-seeds *Coreopsis* spp.

Tidy Tips *Layia platyglossa*

Tolmie Star-tulip *Calochortus tolmiei*

Tolmie's Saxifrage *Saxifraga tolmei*

Tongue-leaf Rainiera *Rainiera stricta*

Toothwort *Lathraea squamaria*

Tragacanth *Astragalus massiliensis*

Tree Spurge *Euphorbia dendroides*

Triglav Hawksbeard *Crepis terglouensis*

Troodos Buttercup *Ranunculus cadmicus var. cyprius*

Troodos Golden Drop *Onosma troodi*

Troodos Orchid *Orchis troodi*

Tuberous Vetch *Lathyrus tuberosus*

Tufted Soapwort *Saponaria caespitosa*

Tufted Vetch *Vicia cracca*

Turban Buttercup *Ranunculus asiaticus*

Twinflower *Linnaea borealis*

Vanilla Leaf *Achlys triphylla*

Vanilla Orchid *Gymnadenia gabasiana*

Violet Cress (also Diamond Flower) *Ionopsidium acaule*

Violet Pasque Flower *Pulsatilla violacea*

Viper's Bugloss *Echium vulgare*

Viper's Grass *Scorzonera humilis*

Vitaliana *Vitaliana primuliflora*

Waldo Gentian *Gentiana setigera*

Wandoo dwarf *Eucalyptus* spp.

Water Avens *Geum rivale*

Wwaterlily Tulip *Tulipa kaufmanniana*

Wendelbo's Corydalis *Corydalis wendelboi*

Western Gorse *Ulex gallii*

Western Hemlock *Tsuga heterophylla*

Western Iberian Paeonies *Paeonia broteroi*

Western Spring Beauty *Claytonia lanceolata*

White Asphodel *Asphodelus albus*

White Avalanche Lily *Erythronium montanum*

White Campion *Silene latifolia*

White Clover *Trifolium repens*

White Cushion Daisy *Celmisia sessilifolius*

White Geranium *Geranium richardsonii*

White Globe Flower *Trollius albiflorus*

White Milk-vetch *Astragalus angustifolius*

White Mountain Heather *Cassiope mertensiana*

White Rock Rose *Helianthemum apenninum*

Whorled Lousewort *Pedicularis verticillata*

Wild Asparagus *Asparagus officinalis*

Wild Carrot *Daucus maritima*

Wild Chamomile *Chamaemelum nobile*

Wild Chives *Allium schoenoprasum*

Wild Daffodils *Narcissus pseudonarcissus*

Wild Pea *Pisum sativum*

Wild Snapdragon *Antirrhinum majus*

Wild Tulip *Tulipa australis*

Winged Broom *Chamaespartium sagittale*

Winter Daffodil *Sternbergia lutea*

Witsenia *Witsenia maura*

Wolfsbane *Aconitum lycoctonum ssp. vulpinum*

Wood Anemone *Anemone nemorosa*

Wood Cranesbill *Geranium sylvaticum*

Wood Pink *Dianthus sylvestris*

Wreath Plant *Lechenaultia macrantha*

Yellow Anemone *Anemone palmata*

Yellow Bee Orchid *Ophrys lutea*

Yellow Bird's Nest *Monotropa hypopitys*

Yellow Coralbells *Elmera racemosa*

Yellow Gazania *Gazania lichtensteinii*

Yellow Gentians *Gentiana lutea*

Yellow Horned Poppy *Glaucium flavum*

Yellow Marsh Orchid *Dactylorhiza romana and similar species*

Yellow Melancholy Thistle *Cirsium erisithales*

Yellow Mountain Daisy *Brachyglottis bellidioides*

Yellow Pheasant's Eye *Adonis vernalis*

Yellow Star-of-Bethlehem *Gagea* spp.

Yellow Wood Anemone *Anemone ranunculoides*

Yellow Woundwort *Stachys recta*